Wilhelm Weber's Main Works on Electrodynamics Translated into English

Volume I: Gauss and Weber's Absolute System of Units

edited by André Koch Torres Assis

Apeiron
Montreal

Published by C. Roy Keys Inc.
4405, rue St-Dominique
Montreal, Quebec H2W 2B2 Canada
http://redshift.vif.com

First Published 2021

Library and Archives Canada Cataloguing in Publication

Title: Wilhelm Weber's main works on electrodynamics translated into English / edited by André Koch Torres Assis.
Names: Weber, Wilhelm Eduard, 1804-1891, author. | Assis, André Koch Torres, 1962- editor.
Description: Includes bibliographical references and indexes. | Contents: Volume I: Gauss and Weber's Absolute System of Units
Identifiers: Canadiana (print) 20210279079 | Canadiana (ebook) 20210279095 | ISBN 9781987980233 (v. 1 ; softcover) | ISBN 9781987980240 (v. 1 ; PDF)
Subjects: LCSH: Electrodynamics.
Classification: LCC QC631 .W4313 2021 | DDC 537.6—dc23

Front cover of Volume I: The picture on the cover of Volume I shows Carl Friedrich Gauss (1777-1855) and his collaborator Wilhelm Weber (1804-1891). Source: Friedrich Zöllner, *Wissenschaftliche Abhandlungen*, Volume 2, Part I (L. Staackmann, Leipzig, 1878), frontispiece.

Wilhelm Weber's Main Works on Electrodynamics Translated into English

Volume I: Gauss and Weber's Absolute System of Units

Λαμπάδια ἔχοντες διαδώσουσιν ἀλλήλοις.
PLATO.

Edited by Andre Koch Torres Assis

Contents

Chapter 1

General Introduction

A. K. T. Assis[1]

Wilhelm Eduard Weber (1804-1891) was one of the main scientists of the XIXth century. From 1831 onwards he worked in close collaboration with Carl Friedrich Gauss (1777-1855). This book contains the English translation of his main works on electrodynamics.

The main goal of these translations is to make Weber's works known to a larger audience, especially those who can not read German. It is addressed to the students and scientists who wish to develop Weber's electrodynamics and applications of his force law to gravitation.

It is also intended for the readers willing to know more about the absolute system of units introduced by Gauss and Weber. Another goal is to allow a better appreciation of the high precision instruments and observational techniques developed by Gauss and Weber during the XIXth century. James Clerk Maxwell (1831-1879), for instance, described in his *Treatise on Electricity and Magnetism* the experimental contributions of Gauss and Weber with the following words:[2]

> The introduction, by W. Weber, of a system of absolute units for the measurement of electrical quantities is one of the most important steps in the progress of the science. Having already, in conjunction with Gauss, placed the measurement of magnetic quantities in the first rank of methods of precision, Weber proceeded in his *Electrodynamic Measurements*[3] not only to lay down sound principles for fixing the units to be employed, but to make determinations of particular electrical quantities in terms of these units, with a degree of accuracy previously unattempted. Both the electromagnetic and the electrostatic systems of units owe their development and practical applications to these researches.

[1]Homepage: www.ifi.unicamp.br/~assis

[2][Max54, Vol. 2, Article 545, pp. 193-194].

[3]Maxwell was referring to a set of Memoirs written by Weber under the general title of *Elektrodynamische Maassbestimmungen*. The title of the Sixth Memoir published in 1871, [Web71], received this translation as *Electrodynamic Measurements* when it was published in 1872, [Web72].

1.1 Weber's Life and Works

Wilhelm Eduard Weber was born in 1804 and died in 1891. He signed his papers as Wilhelm Weber. Weber's complete works were published in six volumes between 1892 and 1894.[4] The earliest obituaries and biographies appeared in 1891-1892.[5] Important ones are those of Heinrich Weber (1839-1928),[6] nephew of Wilhelm Weber and editor of some volumes of his collected works; and Eduard Riecke (1845-1915),[7] who became Weber's successor at Göttingen's University. The most complete biographical studies are those of Widerkehr.[8]

Important works about his life, his collaboration with Gauss, his instruments and measurements, his electrodynamic theory and his scientific activities in general have been published by several authors.[9]

Some experimental techniques utilized by Gauss and Weber were discussed by Friedrich Wilhelm Georg Kohlrausch (1840-1910), the son of Weber's collaborator Rudolf Hermann Arndt Kohlrausch (1809-1858).[10]

There are some excellent homepages with a huge material about Ampère and Gauss which have a direct connection with Weber's electrodynamics.[11] Many publications on Ampère, Gauss and Weber were published in the *Mitteilungen der Gauss-Gesellschaft*, related to the Gauss Society in Göttingen.

1.2 Weber's *Elektrodynamische Maassbestimmungen*

Weber wrote eight major Memoirs under the general title *Elektrodynamische Maassbestimmungen*.[12] This title has received different English translations:

- electrodynamic measurements,[13]

[4] [Web92e], [Web92d], [Web93b], [Web94d], [WW93] and [Web94c].

[5] [F.91], [Ano91] and [Ano92].

[6] [Web92a], [Web92b], [Web92c] and [Web93a].

[7] [Rie92].

[8] Especially: [Wie60] and [Wie67]. See also [Wie64], [Wie73], [Wie82], [Wie88], [Wie90], [Wie91], [Wie92], [Wie93c], [Wie93b], [Wie93a], [Wie94], [Wie97], [Wie04], [Wie07] and [Wie08].

[9] [Dor07], [Sch36], [Hes55], [Kir56] with English translation in [Kir57], [Ros56], [Hes61], [Woo62], [O'R65, volume 2, Chapter 11], [Woo68], [Bie71], [Mol72], [Mil72], [Whi73, Chapter 7], [Cla76], [Rei77], [Can78], [Woo81], [Wis81], [D'A81], [Ros81], [Ass92], [Har82, pp. 32, 96 and 103–107], [Fri82], [Bev83], [Bev84], [Buc84], [Buc85], [JM86], [Ath89], [Ath89], [Arc89], [Mey90], [Mar90], [Sch93b], [Sch93a], [Wie93b], [Bev93], [Dar93a], [Dar93b], [Ass94], [Bev94], [Gra94], [Ass95], [Bev95], [Ten96], [Hec96b], [Hec96a], [D'A96], [Dar96], [Ole96], [GG96], [Ten97b], [Ten97a], [Hec97], [Gib97], [Ass98], [BA98], [Ass99a], [Ass99b], [Glu99], [Hec00], [D'A00], [Dar00], [BA01], [Gar01], [Hec01], [Kär02], [Lin05], [Wol05], [Tim05], [Pic06], [Lin06], [Men06], [AH07], [Hec07b], [Hec07a], [AH09], [Kra10], [AC11], [AWW11], [Hee11], [Rei11], [Kra12], [RR12], [AH13], [Ass13], [Rei13], [Ass14a], [AWW14], [RR14], [Rib14], [Bev14], [GT14], [Ass15], [BA15], [AC15], [Men15], [JM17], [RR17], [AWW18], [Cah18], [Buc20], [Fer20], [Tom20], [BW21b] with English translation in [BW21a], [Wis21], [Hun21] etc.

[10] [Koh83] and [Koh10].

[11] *Ampère et l'Histoire de l'Électricité*: `www.ampere.cnrs.fr`; *Gauss Society Göttingen*: `http://www.gauss-gesellschaft-goettingen.de/gauss-e.htm` and *The Complete Correspondence of Carl Friedrich Gauss*: `https://gauss.adw-goe.de`. See also [Blo05], [Gaua] and [Gaub].

[12] [Web46], [Web52b], [Web52a], [KW57], [Web64], [Web71], [Web78] and [Web94a].

[13] [Web71] with English translation in [Web72], [Kir49, p. 510] with English translation in [Kir50, p. 465], [Web53, p. 163] and [Web66d, p. 163], [Max73, Vol. 2, Article 545, p. 179] and [Max54, Vol. 2, Article 545, pp. 193-194], [Ano92], [Kir56] with English translation in [Kir57, pp. 623 and 625], [Hec07b] and [Wis21, p. 36].

- on the measurement of electro-dynamic forces,[14]
- electrodynamic determinations,[15]
- determinations of electrodynamic measure,[16]
- electrodynamic determinations of measure,[17]
- electrodynamic measure determinations,[18]
- determinations of electrodynamic units.[19]

In French it was translated as "Mesures électrodynamiques".[20]

The word "Maass", nowadays written as "Maß", can be translated as: measurement, measure, dimension, standard etc.

The word "bestimmungen" can be translated as: determinations, regulations, stipulations, specifications etc.

The word "Maassbestimmungen" can be translated as: measurements, measure determinations, determinations of measure etc.

In this book we decided to adopt the translation of *Elektrodynamische Maassbestimmungen* as that utilized during Weber's lifetime, namely, *Electrodynamic Measurements*. His sixth Memoir had been published in 1871: *Elektrodynamische Maassbestimmungen insbesondere über das Princip der Erhaltung der Energie*.[21] It was translated by George Carey Foster (1835-1919) and published in the *Philosophical Magazine* of 1872: *Electrodynamic measurements — Sixth memoir, relating specially to the principle of the conservation of energy*.[22] In this book we will also specify in the title of each of the 8 major Memoir to which one it refers to, just as was made by Foster in the title of the English translation of this sixth Memoir.

1.3 Practical Aspects of the Project

Beyond Weber's main papers on electrodynamics, some other translations by scientists related to Weber's works were also included in this project. There are papers by Carl Friedrich Gauss, Gustav Kirchhoff (1824-1887), Johann Christian Poggendorff (1796-1877), Carl Gottfried Neumann (1832-1925) and François Félix Tisserand (1845-1896). They were included due to the mutual impact between their researches and those of Wilhelm Weber. It is then possible to have a better idea of the development of Weber's law applied to electrodynamics and gravitation during the XIXth century.

Two groups of translations were included in this project. The first one were those published during the XIXth century. In this case I typed myself all papers in the text editor

[14][Web48a] with English translation in [Web52d], [Web66f] and [Web19].

[15][Kir57b, p. 194] with English translation in [Kir57a, p. 394].

[16][JM86, Vol. I, p. 140] and [JM17, p. 159], [Web46] with English translation in [Web07], [Web94b] with English translation in [Web08] and [Wis21, p. 36].

[17][Hec96b, pp. 31-32].

[18][Dar00, p. 56].

[19][Jac06, p. 113].

[20][Web46] with partial French translation in [Web87].

[21][Web71].

[22][Web72].

LaTeX. When necessary I made some modifications in these translations by comparing them with the original German texts. I pointed out the relevant changes in footnotes.

The second group is related to translations made from the 1990's onwards. In this last case I had a direct personal contact, at least by e-mail, either with the translators or with the editors of the translations. I received most of the translations of this second group typed in the text editor Word. In these cases I first converted the text content into LaTeX. I then typed in LaTeX the formulas and mathematical symbols in the middle of the sentences, the mathematical equations, the numerical tables and inserted all figures, footnotes and references. I checked the English translation and suggested some modifications. Only after myself and the translator (and/or editor) reached an agreement on the final version, was it posted online in my homepage in PDF format. The same happened when I received the translations typed directly into LaTeX. The translations presented in this book are the latest versions. There are improvements in the footnotes etc. relative to the versions posted in my homepage.

The words between square brackets, [ ], were introduced by myself or by the translators in order to clarify the meaning of some sentences.

I inserted footnotes with similar content in different Chapters. This was done on purpose. This information will then be available to readers interested in any specific work which can be read independently from the other Chapters.

A great effort has been spent trying to locate and list relevant references which were mentioned directly or indirectly by Weber. Sometimes he quoted a single name. I then tried to locate whom this person might had been and which papers of this scientist might Weber be referring to. This information was gathered on the footnotes contained in each translation. Whenever possible I tried to insert complete references with the full name of the Journal, full title of the paper, first and last pages etc. I also tried to quote all translations known to me of any specific reference in different languages. These references help to contextualize Weber's influences, methods, contacts and researches.

Obviously all remaining mistakes in both groups of papers are my own responsibility, as I typed almost everything into LaTeX. Errors are unavoidable in such a huge project. Comments and suggestions for improvement are most welcome. I would also like to continue working on this project by making available English translations of other papers by Weber which have not been contemplated in the present edition. People interested in translating other works along these lines can reach me easily by e-mail.

I hope these English translations of Weber's main works on electrodynamics will help to bring his amazing theory and experiments to the knowledge of a larger audience. It can then be further developed and brought once more to the forefront of modern science.

1.4 Acknowledgments

My gratitude goes mainly to my colleagues who understood the importance of this project, made the translations and helped to edit the works presented in this book: Laurence Hecht, David H. Delphenich, Urs Frauenfelder, Joa Weber, Peter Marquardt, Hermann Härtel, Jonathan Tennenbaum and Peyman Ghaffari. Without their support this book would never be published.

I thank 21st Century Science Associates and its Editor-in-Chief Jason Ross for the permission to include in this book the English translations of the papers by Gauss and Weber which were published in 21st Century Magazine and in their homepage.[23]

I would like to thank as well several other colleagues for their suggestions, references, ideas, support and encouragement: Karin Reich, Gudrun Wolfschmidt, Martin Tajmar, Elena Roussanova, Simon Maher, Christof Baumgärtel, Frederick David Tombe, Elisabeth Becker-Schmollmann, Gilberto Orengo, Decio Schaffer, Robert W. Gray, Alan Aversa, Alexander Unzicker, Kjell Prytz, Jan Rak, Karel Janecek, Reiner Ziefle, Wallace Thornhill, Tim Hooker, Lucy Wyatt, Thomas Herb, David de Hilster, John Lord, Karl-Heinz Glassmeier, Mathias Hüfner, Orges Leka, João Paulo Martins de Castro Chaib, Fabio Menezes de Souza Lima, Paulo Henrique Dias Menezes, Arthur Baraov, Pietro Cerreta, Riccardo Urigu, Chuck Stevens, Wallace do Couto Boaventura, Danny Augusto Vieira Tonidandel, Marcos Cesar Danhoni Neves, Daniel Gardelli, Domingos Soares, Arden Zylberstajn, Fernando Lang da Silveira, Lúcio Costa, Breno Arsioli Moura, José Emílio Maiorino, Elizabeth Silber, Hannes Täger, Amitabha Ghosh, Julian Barbour, Christine Blondel, Bertrand Wolff, Ana Paula Bispo da Silva, Daniel dos Anjos Silva, Frederico Ayres de Oliveira Neto, Mario Novello, Haroldo Fraga Campos Velho, Kathryn Olesko, Yuri Heymann, Joerg Fischera, Juan Muñoz Madrid, David Bower, Ismo V. Lindell, Ovidio Bucci, John Plaice, Klaus Hentschel, Tony C. Scott, Markus Wirz, John Eastmond, José Manuel Ferreirós Dominguez, Bernard Guy, Koen van Vlaenderen, Kirk McDonald, Rolf Laeuppi, Franz Pichler, Samer al Duleimi, Greg Volk, Mario Wingert, Michael D. Godfrey, Jocelyne Lopez, David Dameron, Mario J. Pinheiro, Hermann Borotschnig, Mischa Moerkamp, Mario Natiello, Júlio Akashi Hernandes, Simon Brunnquell, Andreas Otte, Max Tran, Konstantinos Kifonidis, Hans Günter Dosch, Albert Gerard Gluckman, Steve Hutcheon, Christian Ucke, Hartwig Thim, Steffen Kühn, Frierich Steinle, Peter Heering, Rainer Müller, Sahand Tokasi, Cesar Pagan, Joachim Schlichting, Matthias Heumesser, Kai Cieliebak, Stefan Suhr, Marcelo Bueno, Dario Thober, Manfred Pohl, Andreas Schuldei, Fritz A. Krafft, Wolfgang Engelhardt, Osvaldo Pessoa Jr., Alvaro Vannucci, Iberê Caldas, Edmundo Capelas de Oliveira, Peter Puschnig, Michael de Carvalho, Pieter Jacqmaer, Matthias Dörries, Helmut Hansen, Juan Manuel Montes Martos, João José Caluzi, Karl-Heinz Schlote, Junichiro Fukai, Wolfgang R. Dick and Neal Graneau.

Roy Keys, the Editor of Apeiron, has been a supporter for many years. Without his encouragement some of my books might not have been published. He was very receptive to this particular project.

I wish to thank the Institute of Physics of the University of Campinas — UNICAMP, which gave the necessary support for undertaking this work. I thank also the Alexander von Humboldt Foundation of Germany for three Research Fellowships which allowed me to study German at the Goethe Institute in Göttingen (from April to July 2001) and to work at the Institute for the History of Natural Sciences of Hamburg University (from August 2001 to November 2002 and from February to May 2009) and at the Dresden University of Technology (from April to June 2014). I made extremely important personal contacts and collected considerable bibliographic material for this project during these research stays in Germany.

[23]`http://21stcenturysciencetech.com` and `http://21sci-tech.com/translation.html`

Chapter 2

Introduction to Volume I

A. K. T. Assis[24]

The picture on the cover of Volume 1 shows Carl Friedrich Gauss (1777-1855) and his collaborator Wilhelm Weber (1804-1891). It comes from the frontispiece of the second volume of a book by Friedrich Zöllner (1834-1882) containing the collection of his papers.[25] Zöllner explained the origins of these pictures and mottos on pages v-vii of his book.[26]

This first Volume begins with Gauss' work on the absolute measure of the Earth's magnetic force. It was announced in 1832 and the full paper in Latin circulated in small edition in 1833. The first complete German translation appeared in 1833, although the original paper in Latin was published only in 1841. In this work Gauss introduced the absolute system of units in which electromagnetic magnitudes were measured based on the dimensions of length, mass and time (specifically millimeter, milligram and second). In particular, he obtained absolute measures of the Earth's magnetic force and of the magnetic moment of a magnetized bar. As he acknowledged in the paper, he was assisted by Weber in many ways. Weber had obtained the professorship of physics at Göttingen University in 1831 under the recommendation of Gauss.

This volume also contains translations of other papers by Gauss, Weber and Wöhler up to 1842. They deal with the Magnetic Association created by Gauss and Weber, their instruments to perform high precision magnetic measurements including the unifilar and bifilar magnetometers, the composition of galvanic piles and the electrochemical equivalent of water. A large portion of Weber's works in physics was to implement and extend the absolute system of units. In particular, he created methods to effectively obtain high precision absolute measures of electric charge, electric current, electromotive force and resistance.

[24] Homepage: `www.ifi.unicamp.br/~assis`

[25] [Zöl78].

[26] See also [Fer07].

Chapter 3

Some Personal Notes from the Translators

3.1 Why Did I Decide to Work in this Project of the English Translation of Weber's Main Works on Electrodynamics

André Koch Torres Assis
Institute of Physics 'Gleb Wataghin'
University of Campinas — UNICAMP
13083-859 Campinas, SP, Brazil
E-mail: assis@ifi.unicamp.br
Homepage: www.ifi.unicamp.br/~assis

I decided to work on this project because I believe in Weber's electrodynamics and want to make it better known to a larger audience.

I discovered Weber's electrodynamics in Whittaker's book *A History of the Theories of Aether and Electricity* which I read for the first time in 1985.[27] Since then I learned many properties of Weber's force which were in agreement with my physical intuition. I also saw how powerful it was. Details and references of what I will say here can be found in Weber's works and in my books quoted below.

Weber published his force law in 1846. He unified Coulomb's force between electrified particles (1785), Ampère's force between current elements (1822 and 1826), and Faraday's law of induction (1831). Ampère's force is a central force, pointing along the straight line connecting the current elements, no matter the directions of the electric currents in these elements. It also complies with Newton's action and reaction law, just like Coulomb's force and Newton's law of gravitation (1687).[28] Weber's force between two electrified particles is always along the straight line connecting them. It complies with Newton's action and reac-

[27] [Whi73].

[28] Isaac Newton (1642-1727). See [New34] and [New99]. Portuguese translation in [New90], [New08] and [New10].

tion law in the strong form. It is a generalization of coulomb's force. It depends not only on the distance r between the interacting particles, but also on their *relative* velocity dr/dt and on their *relative* acceleration d^2r/dt^2. It can be deduced from a velocity dependent potential energy which was also introduced by Weber in 1848. It complies with conservation of linear momentum, angular momentum and energy. The whole of electrostatics (Coulomb's force and Gauss' law) is contained in Weber's law. Ampère's force between current elements can also be deduced from Weber's law. The circuital magnetic law can also be deduced from Weber's force, including the displacement current. Weber succeeded in deducing Faraday's law of induction from his force. The first quantitative connection between optics and electrodynamics originated also in Weber's electrodynamics. This was the result of Weber and Kohlrausch's measurement of a fundamental constant which Weber had introduced in his force. The result of their measurement was published in 1855, 1856 and 1857. In 1857 Kirchhoff and Weber deduced independently from one another the complete telegraph equation by taking into account (in modern terms) not only the resistance and capacitance of the wire, but especially its self-inductance. Both of them worked with Weber's electrodynamics. Their works were published in 1857 and 1864. They showed, in particular, that when the resistance of the wire was negligible, the telegraph equation reduced to the wave equation. The velocity of propagation of an electric wave along the wire was then shown to be independent of the cross section of the wire, of its conductivity and of the density of electricity along the surface of the wire. Its value was equal to the known light velocity in vacuum. This remarkable result of Weber's electrodynamics indicated for the first time in the history of physics a direct and quantitative connection between electrodynamics and optics.

During my undergraduate and graduate studies in physics I was introduced to the so-called Lorentz force. As usually mentioned in the textbooks, if an electrified particle with charge q is moving with velocity $\vec{v}$ in the presence of an electric field $\vec{E}$ and a magnetic field $\vec{B}$, then the Lorentz force $\vec{F}$ acting on this charge is given by $\vec{F} = q\vec{E} + q\vec{v} \times \vec{B}$. The textbooks usually do not specify the meaning of this velocity. Is it the velocity of the charge q relative to what? This force can only be applied or utilized when we know the meaning of this velocity. When I discovered that nowadays this velocity $\vec{v}$ is interpreted as the velocity of the charge q relative to the observer, I did not accept it. This interpretation was against my physical intuition, after all the charge q is not interacting with the observer. It is interacting with other electrified particles.

Let me clarify my point of view with an analogy. Consider a scientist on Earth studying the orbit of a satellite of Jupiter. According to Newton's law of gravitation, in the analysis of this problem the relevant parameter specifying the orbit and its properties is the distance r between the the center of the satellite and the center of Jupiter, and not the distance between the satellite and the observer on Earth. In the future we may conclude that the gravitational force depends not only on the position of the bodies, but also on their velocities. In this specific example, I believe that the orbit of the satellite and its properties would then depend on its velocity relative to Jupiter, dr/dt, but not on its velocity relative to the observer on Earth.

Weber's law called my attention due to its philosophically appealing properties, namely, it acts along the straight line connecting the interacting particles and it complies with Newton's action and reaction law. Moreover, it depends only on the distance r between the interacting particles, their *relative* radial velocity dr/dt and their *relative* radial acceleration d^2r/dt^2. These are intrinsic properties of the system. The observer (or frame of reference) does not matter for the values of r, dr/dt and d^2r/dt^2. These magnitudes have the same values in all frames of reference, even for non-inertial reference frames. Later on I called them *relational magnitudes*.

Further discussions of these topics can be found in Sections 3.1 (Multiple Definitions of the Field Concept), 3.2 (These Different Field Definitions Contradict One Another), 15.5 (Origins and Meanings of the Velocity $\vec{v}$ which Appears in the Magnetic Force $q\vec{v}\times\vec{B}$), and in Appendix A (Relational Magnitudes) of the book *Relational Mechanics and Implementation of Mach's Principle with Weber's Gravitational Force.*[29] In Section 15.5 I discuss the four different definitions of the velocity $\vec{v}$ utilized in the magnetic force as given by (a) J. C. Maxwell; (b) J. J. Thomson and O. Heaviside; (c) H. A. Lorentz; and (d) A. Einstein. These definitions are different from one another. Therefore, although the mathematical expression of this force may be the same for these authors, their force laws are also different from each other. These forces do not belong to the same theory, but to four different theories.

My initial approach was to extend Weber's law to gravitation. That is, to suppose that Newton's force of gravitation should be complemented by a term depending on the relative velocity between the interacting masses and another term depending on the relative acceleration between them. I only began to work seriously with Weber's law at the end of the 1980's when I succeeded in implementing mathematically Mach's principle with Weber's force applied to gravitation.[30] *It was then possible to show that kinematically equivalent motions were also dynamically equivalent*, a very intuitive result. Newton's bucket experiment could be explained not only with the bucket and water spinning together relative to a stationary background of distant matter, but also with the bucket and water stationary, while the distant bodies in the universe were spinning together around the axis of the bucket. Both points of view led to the same curvature of the water in the bucket, provided the relative rotation between the bucket and the set of distant bodies was the same in both cases. Likewise Foucault's pendulum experiment could be explained not only with the diurnal rotation of the Earth relative to the stationary background of distant matter, but also with a stationary Earth, while the distant bodies in the universe were spinning daily around the Earth's axis. *It was also possible to deduce Newton's second law of motion from Weber's law applied to gravitation, coupled with the principle of dynamical equilibrium.* According to this principle, the sum of all forces acting on any body is always zero in all frames of reference.

Since my first paper in 1989, I published several books dealing directly with Ampère's electrodynamics and Weber's law applied to electromagnetism and gravitation:

- Weber's Electrodynamics.[31]
- Relational Mechanics.[32]
- Inductance and Force Calculations in Electrical Circuits.[33]
- The Electric Force of a Current: Weber and the Surface Charges of Resistive Conductors Carrying Steady Currents.[34]

[29] In English: [Ass14a]. In Portuguese: [Ass13].
[30] [Ass89].
[31] In English: [Ass94]. In Portuguese: [Ass92], [Ass95] and [Ass15].
[32] In English: [Ass99a]. In Portuguese: [Ass98] and [Ass99b].
[33] In English: [BA01]. In Portuguese: [BA98] and [BA15].
[34] In English: [AH07]. In Portuguese: [AH09]. In German: [AH13].

- Ampère's Electrodynamics — Analysis of the Meaning and Evolution of Ampère's Force between Current Elements, together with a Complete Translation of His Masterpiece: Theory of Electrodynamic Phenomena, Uniquely Deduced from Experience.[35]

- Weber's Planetary Model of the Atom.[36]

- Relational Mechanics and Implementation of Mach's Principle with Weber's Gravitational Force.[37]

My papers on these subjects can be found in my homepage.

I always had a great interest in the history of science and I wished to read Weber's original works. But at that time I still did not know German. For many years I had to rely on English translations of Weber's works of 1848 and 1871,[38] on English translations of two papers of 1857 by Kirchhoff on the propagation of electromagnetic signals along conducting wires utilizing Weber's law,[39] and on secondary sources by other authors. The second paper by Kirchhoff was translated at my request by the late Peter Graneau (1921-2014) with whom I had worked for one year in Boston, USA, from 1991 to 1992, supported by a research fellowship given by FAPESP (São Paulo Research Foundation, Brazil).[40]

In 2000 I received the invitation from the late Karl-Heinrich Wiederkehr (1922-2012) and Karin Reich to work at the Institute for the History of Natural Sciences of Hamburg University.[41] Herr Wiederkehr had written the main biography of Weber.[42] I then began to study German for one semester in my University. I received a fellowship from the Alexander von Humboldt Foundation of Germany in order to develop the project "Weber's Law Applied to Electromagnetism and Gravitation". Before the beginning of the project, Humboldt Foundation gave me the opportunity to take part on an intensive 4 months course of German at the Goethe Institute in Göttingen. It was fascinating to finally learn German, especially studying at the city where Gauss and Weber spent most of their careers. I worked with K. H. Wiederkehr and K. Reich in Hamburg from August 2001 to November 2002. During this period I read most of the 6 volumes of Weber's collected works. One of the jewels I discovered was his pioneering calculation of the distribution of charges spread along the surface of resistive conductors carrying steady currents, a subject on which I had been working for several years. In 2007 I published in collaboration with J. A. Hernandes a book discussing Weber's calculations.[43]

It was during this period of 2001-2002 while in Hamburg that I first had the idea of this project of an English translations of Weber's main works on electrodynamics. I knew that many students and scientists around the world could not read German and would be in the same situation that I had experienced for many years. It seemed to me that the neglect of Weber's electrodynamics during the whole of the XXth century was largely due to the lack

[35]In English: [AC15]. In Portuguese: [AC11].

[36]In English: [AWW11]. In Portuguese: [AWW14]. In German: [AWW18].

[37]In English: [Ass14a]. In Portuguese: [Ass13].

[38][Web48a] with English translation in [Web52d], [Web66f] and [Web19]; and [Web71] with English translation in [Web72].

[39][Kir57b] and [Kir57c] with English translations in [Kir57a] and [GA94], respectively.

[40][AG95], [AG96], see also [Ass14b].

[41][ARW02], [AW03] and [ARW04].

[42][Wie60] and [Wie67].

[43][AH07] with Portuguese translation in [AH09] and German translation in [AH13].

of knowledge about his original works. An English translation of his main works might help enormously to bring his ideas once more to the forefront of modern science.

In this period I also met personally Laurence Hecht at Göttingen University Faculty of Physics when we both were given a tour by Professor G. Beuermann of the rooms where some of the original equipment used by Gauss and Weber in their electrodynamic experiments are preserved.[44] I helped him to edit the first English translation of Gauss' fundamental work introducing the absolute system of units in magnetism.[45] I also helped him to edit and publish Weber and Kohlrausch's 1856 paper on the amount of electricity which flows through the cross-section of the circuit in galvanic currents.[46] Both of these papers had been translated by the late S. P. Johnson.[47] I also helped him to edit Weber's first and last major Memoirs on *Elektrodynamische Maassbestimmungen* during 2007-2008 and the project finally began.[48]

I returned to Hamburg University once more from February to May 2009. This time I worked with K. H. Wiederkehr and Gudrun Wolfschmidt on the project "Weber's Planetary Model of the Atom" which gave rise to a book with the same title.[49]

From April to June 2014, I worked with Martin Tajmar at Dresden University of Technology on the project "Exploring the Effective Inertial Mass of Particles".[50]

In all these occasions I was supported by the Alexander von Humboldt Foundation of Germany with Humboldt Research Fellowships. I was always extremely well received, had all the necessary scientific support from these Universities, made many friends and important personal contacts. During these three research periods in Germany I could read and collect a huge amount of material which has been essential for the development of this project.

I have always been involved in many professional activities simultaneously. Only from 2018 onwards could I devote most of my time to this project of the English translation of Weber's main works on electrodynamics. Happily I could find many competent colleagues who could see the importance of these translations and helped me to make it succeed.

Weber's electrodynamics disappeared from the textbooks and from the consciousness of the scientific community during the whole XXth century. I was happy to discover it in Whittaker's book. I was also very fortunate when I decided to work with it as applied to electromagnetism and gravitation. This conscious choice was a turning point in my scientific career. It made all the difference in my professional life. Since then I have been fighting every single day against the establishment and in favour of my physical intuitions. I have been very satisfied with my decision. I feel myself completely realized scientifically and believe that I accomplished much more than I ever dreamt of as a student.

With the help of my colleagues, I now share with all of you Weber's works translated into English. You can then learn Weber's electrodynamics from his own words. I hope that by seeing the great importance of his works, you will also follow along his footsteps, writing research papers and textbooks, teaching it to your students, extending and developing his theory and experiments as applied to electromagnetism and gravitation.

---

[44] See `https://www.uni-goettingen.de/en/47114.html` and `http://physicalisches-cabinet.uni-goettingen.de`. See also [Beu d] and [Beu97].

[45] [Gau94] with English translation in [Gau03] and Portuguese translation in [Ass03b].

[46] [WK56] with English translation in [WK03] and Portuguese translation in [WK08]. See also [Ass03a].

[47] [Joh97].

[48] [Web46] with a partial French translation in [Web87] and a complete English translation in [Web07]; and [Web94b] with English translation in [Web08].

[49] [AWW11] with Portuguese translation in [AWW14] and German translation in [AWW18].

[50] [TA15b] with German translation in [TA15c], [TA15a], [TA16], [AT17] and [AT19].

3.2 Why to Work for the Spreading of Weber's Ideas

Hermann Härtel
Guest scientist at
ITAP - Institute for Theoretical Physics and Astrophysics
University Kiel, Germany
E-mail: haertel@astrophysik.uni-kiel.de
`http://www.astrophysik.uni-kiel.de/~hhaertel/index_e.htm`

During all the time when studying physics and later when trying to improve the teaching of physics, I struggled with the topic "electromagnetic induction", based on Faraday's law and the Lorentz force. I never really understood this topic, I finally accepted it — with reluctance and displeasure. It made me feel more and more uncomfortable over the years, especially at times when I was supposed to teach it to students and once again found out from the achieved learning outcomes that I had failed.

Explanation of induction in the traditional way does not create clarity, nor does it give the impression of being understood. The essential learning success is normally to be able to predict the direction of the induced current using the right-hand rule.

Now in my old age I started working on this material again, inspired by a reference to the works of Wilhelm Weber published in 1846. My work on this topic was initially based on the question: Could it be that there is an alternative to a knowledge such as Faraday's law and the Lorentz force, which has been known and universally recognized for more than 150 years? And should this alternative even have certain advantages, above all didactic ones?

To my astonishment, my initial skepticism waned to the extent that I realized how all known induction phenomena could be explained in a uniform and transparent way based on Weber's approach.

To change long lasting knowledge like Faraday's flux law and Lorentz approach will certainly take a long time. But being retired and without any other obligations, I am glad to have the opportunity to help spreading Weber's ideas.

3.3 How We Got Interested in Weber's Electrodynamics

Urs Frauenfelder[1] and Joa Weber[2]

1 - Mathematisches Institut
Universität Augsburg
Augsburg, Germany
E-mail: urs.frauenfelder@math.uni-augsburg.de

2 - Institute of Mathematics
University of Campinas — UNICAMP
13083-859 Campinas, SP, Brazil
E-mails: joa@ime.unicamp.br and joa@math.uni-bielefeld.de
Homepage: `https://freedom-and-science.neocities.org`

We are working in Symplectic geometry. This is the geometry which lies behind Hamiltonian dynamics in particular Celestial mechanics. Our research topic in Symplectic geometry is Floer homology. An important ingredient in Floer homology is that physical orbits can be described variationally as critical points of an action functional. This is also known as principle of least action although this name is a bit misleading, since in many cases the physical orbits are not minima of the action but instead of that saddle points. Now there is a fascinating interplay between critical points and topology. You can see this already in finite dimensions. If you look at a sphere in front of you, the height has a maximum and a minimum but no saddle point. If instead of that you look at a tyre or more mathematically speaking a torus, you see not only a maximum and a minimum of the height but also two saddle points. The difference is that the tyre has a hole but the sphere has no one. If you have more holes the topology gets more complicated and this forces more saddle points, which for the action functionals in classical mechanics correspond to physical solutions. Floer homology studies this connection between the number of holes and dynamics.

Urs Frauenfelder was researching on the question what happens to Floer homology, when you consider functionals with delay. This happens if the particles do not interact instantaneously but with some retardation. He was collaborating with Joa Weber on this question and we wanted to know historically in which context such functionals showed up. In this way he discovered the paper from 1868 of Carl Neumann "Die Prinzipien der Elektrodynamik", in which Carl Neumann derived Weber's law from a delayed functional.[51] It was not only an amusing coincidence that Joa Weber had the same family name as Wilhelm Weber, although he does not seem to be related to him, but also that Joa Weber is working at the University of Campinas as André. This was in 2019 and the three of us had the idea to organize an *Advanced School* at UNICAMP bringing together mathematicians and physicists.[52] The

[51] [Neu68] with English translation in [Neu20].

[52] `https://freedom-and-science.neocities.org/M/20-WED/WED.html`

public interest in the *recorded videos* is overwhelming.[53]

The first time Urs Frauenfelder heard of Weber's electrodynamics was when he still was a student at ETH in Zurich and Peter Graneau gave a talk on "Newtonian Electrodynamics".[54] Through personal discussions with André and his books we got a much clearer insight into the tragic history of Wilhelm Weber's electrodynamics. Although our own research is motivated by completely different questions then the ones of Wilhelm Weber, his work is still inspiring us and we are convinced that through its study we can get a deeper understanding of our world full of riddles.

[53] `https://www.youtube.com/channel/UCOIeUkMqXstDrJKAsn11UgA/videos?view=0&sort=p`

[54] See, for instance, [GG96].

3.4 Participating in the *Weber* Project

Peter Marquardt
E-mail: marquardtp@gmail.com

In the scene of non conformists I am a late comer. *Paul Wesley* (1921 - 2007) was my ticket to the world of serious critics concerning physics in general and electromagnetism in particular.[55] During the twelve years I exchanged and shared dissident views of physics with him, he became my favorite physics teacher. *George Galeczki* (1945 - 2016),[56] working at the same institute for experimental physics as I during the late 80ies, was my first source to vast material disproving special relativity, quickly convincing me to join the dissidents. *George* had been in contact with *Paul Wesley* before and encouraged me to attend a private meeting organized by *Halton Arp* in Munich.[57] After meeting *Paul* there I was introduced to a world of critical thinkers hitherto unknown to me. Some of these have been active for a long time before me. Owing to their previous work I am the lucky one to find himself on a well prepared road for adopting a distant view of physics.

In the years to come I attended a couple of workshops in Cologne (organized by *George*), in Lanzarote in 2002 (my first encounter with *André Assis*) and in the USA (organized by the then Natural Philosophy Alliance). I unlearned more of orthodox physics, among which the one and only officially accepted electrodynamics (*Maxwell*'s) that eventually was to pave the path to special relativity.

Mainly from *Paul* I leaned that *Maxwell*, the one kind of electromagnetism in official use, is defective in many ways in spite of its impressive performance. It owes its fame to usual dogmas like exclusively admitting transverse (= *Lorentz*) forces and transverse electromagnetic waves. Based on two faulty assumptions (*Faraday*'s induction and the *Biot-Savart* force, theories lacking consistence and unsuited for radiation), the *Maxwell* success to yield waves must be attributed to mathematical manipulation, not to a physical basis. When the sources (needed to define the E and B fields) are dismissed for free space we are left with a situation similar to the grin of *Lewis Carroll*'s Cheshire Cat when the cat is gone. The grin is enough for mathematicians to carry on, a valuable lesson teaching us that a correct result is by no means sufficient to prove a theory right.

Critical inspection of the sources prior to *Maxwell* reveal more consistent but vastly neglected models for interacting charges. Two main actors, practically ignored by mainstream scientists, deserve special attention: *André Marie Ampère* (whom *Maxwell* called the "Newton of electricity")[58] and *Wilhelm Eduard Weber*, both underrated and hence under-represented in the standard literature.

Ampère plays second fiddle in the orchestra of physics (if the establishment lets him play at all). His longitudinal forces, although experimentally established,[59] are excluded from mainstream textbooks and lectures. Same with *Weber*'s merits - in my student time, lectures on electromagnetism never mentioned *Weber*, not even for his determination of c together with *Kohlrausch*. Just the SI unit named after him survived, 1 Wb = 1 Vs. *Weber*'s work had been neglected by the establishment mainly because his theory did not deliver waves. It

[55] See `https://www.jamespaulwesley.org`.
[56] See [Gal93] and [GM95].
[57] See [Arp87], [Arp98] with Portuguese translation in [Arp01]. See also `https://www.haltonarp.com`.
[58] [Max54, Vol. 2, Article 528, page 175].
[59] [Gra94].

can, however, be extended to a field theory with retardation and radiation covering *Maxwell*'s as a limited special case and it can be shown to yield *Ampère*'s force.[60] This brings us to a real gem we find among *Weber*'s prophetic ideas: He may be considered the father of the dynamic (i.e. velocity dependent) potential that enriches the static *Cavendish-Coulomb* potential by an additional term $V^2/2c^2$ with far reaching consequences: This generalization gives a $(velocity)^2$ term its own life as potential, including our ubiquitous friend, the notorious c^2. *Weber*'s dynamic potential is a historical action-at-a-distance model, also applicable to gravitation complying with *Mach*'s Principle.[61]

What about *Weber* and c^2? The most famous formula of all science, attributed to you-know-who, can be traced back to *Weber*'s writings.[62] It is especially this trophy that had intrigued me. That famous c^2 invades so many formulae and topics of physics before 1905. Yet it is officially regarded as a medal of special relativity, but, in fact, disproves the dogma of the observer related constant c. *Paul Wesley* refers to c^2 as "Weber's cosmological condition",[63] the physical background of the unique global reference for light propagation that defines the only inertial system we have. There is indeed a lot about *Weber* that waits to be discovered or rediscovered.

With his importance both in experiment and theory and with his contributions to measurements and units (cgs, still used by some) *Weber* was a multi talented all round physicist. It is a matter of scientific honesty to put the spotlight on him in order to get (or regain) the official reception he deserves. Making *Weber* accessible to a broad community is a must.

In early 2020, *André Assis* had asked me to participate in the translators' job. I felt honored, but looking through the pages of "my" part written in typical 19th language, somewhat strange even for Germans, I was almost discouraged. *Weber*, a child of his time, wrote in a somewhat circumstantial and cryptic style. Those mile long sentences looked like more than just the challenge to unravel their grammar. Moreover, the translation was to stick as closely to the original language in order to convey as much of *Weber* as possible. But then, seeing there was already so much work done in this huge project by other translators, I would feel guilty not making an effort; so I joined the project... Eventually it proved to be an exciting and most rewarding adventure.

The present choice from the phenomena investigated by *Weber*, oscillations, were to become my part in the great adventure of preparing his writings for an international public.

Considering *Weber*'s essay "Elektrodynamische Maassbestimmungen insbesondere über elektrische Schwingungen",[64] we may rightfully ask whether *Tesla*'s pioneering work was possible without *Weber*'s pioneering work. Likewise, his co-operation with *Gauss* on the telegraph opened the door to a new era of technology including hitherto unknown effects such as the skin effect.

The more I went into my part of the task the more it fascinated me. I'm positive I share this fascination with all others who translated and with all who will have access to *Weber*'s works on a more international stage. It provides a welcome invitation to get back to the sources for the sake of critical thinking.

Cologne, April 2021

---

[60] [Wes91] and [WM06].

[61] [Ass89] and [Wes91].

[62] [GG93].

[63] [Wes01].

[64] [Web64] with English translation in [Web21a].

3.5 Buried Treasure in Weber's Research

David H. Delphenich
E-mail: feedback@neo-classical-physics.info
Homepage: `www.neo-classical-physics.info/index.html`

On the surface of things, the series of articles by Wilhelm Weber on electrodynamical measurements would appear to have a largely historical significance to them. After all, they are primarily concerned with Nineteenth Century experimental technology and laboratory techniques. As such, one would expect that the advancement of experimental technology would render the discussion essentially obsolete at this point in the history of physics.

However, it is in the discussion of the theoretical basis for the experiments that Weber discusses a problem that rarely gets posed in the current era, and with rather intriguing results. That is the problem of how one would correct Coulomb's law of electrostatics for the relative motion of the interacting charges.

In its static form, that law expresses the force **F** of mutual attraction or repulsion of two electric charges q_1 and q_2 at points P_1 and P_2 in space that are separated by a distance of r in the form:

$$\mathbf{F}(P_1, P_2) = \frac{q_1 q_2}{r^2}\hat{\mathbf{r}}(P_1, P_2) ,$$

in which $\hat{\mathbf{r}}(P_1, P_2)$ represents the unit vector that points from P_1 to P_2. However, since the law pertains to only electrostatics, there is an implicit constraint imposed upon the law, namely, that the distance r must be constant in time, which is essentially a rigidity constraint.

Although one can configure experimental arrangements in which the interacting charges are forcibly constrained to remain at fixed points in space, nonetheless, that scenario does not exhaust the set of experimentally-realizable possibilities. Indeed, the scattering of charges by other charges is a common situation in which one must consider the relative motion of charges that are otherwise free to move in space. Thus, the question arises of whether the force of mutual attraction or repulsion is not just a function of the distance between the charges, but also their state of relative motion, i.e., position, velocity, acceleration, etc. Interestingly, that problem of electrodynamics never seems to be posed in modern discussions of classical electromagnetism, which simply goes directly from electrostatics to magnetostatics in most cases.

Intuitively, one knows that the relative motion of charges will generate magnetic fields, which will exert forces on currents (i.e., moving charges). Hence, there is a ring of plausibility to the idea that perhaps Coulomb's law is fundamentally incomplete in that it does not include the state of relative motion of the interacting charges. Here is where Weber takes one step further and derives an expression for the force of mutual attraction/repulsion that includes contributions from the relative velocity **v** and acceleration **a**, namely, he derives the following expression for the magnitude of that force in terms of the magnitudes v_0 and a of **v** and **a**, respectively:[65]

[65]Although Weber did not originally identify the coefficients in his expression with $c^2/2$ and c^2, where c

$$F(r, v, a) = \frac{q_1 q_2}{r^2}\left(1 - \frac{1}{2}\frac{v_0^2}{c^2} + \frac{ar}{c^2}\right) .$$

Here is where things get intriguing. If one recalls the formula from elementary kinematics that relates the change in speed over distance when one assumes that acceleration is constant, namely:

$$v^2 = v_0^2 + 2ar ,$$

where v_0 is the initial speed and v is the speed after the distance r has been traversed, then one sees that if one assumes that the acceleration a is actually negative, then the term in parentheses above will take the form:

$$1 - \frac{1}{2}\frac{v^2}{c^2} .$$

Now, compare that to the expression for the distance r when one introduces the correction from special relativity that accounts for the apparent contraction of length under relative motion:

$$r\left(1 - \frac{v^2}{c^2}\right)^{-1/2} .$$

That would imply that $1/r^2$ would have to be corrected to:

$$\frac{1}{r^2}\left(1 - \frac{v^2}{c^2}\right) ,$$

which differs from Weber's expression only by the somewhat-peculiar replacement of c with $c/\sqrt{2}$.

Of course, Weber's work predated the theory of special relativity by several decades, so there is something prescient about the result above in that sense. Furthermore, it had the intriguing consequence that since the correction term can vanish, and in fact, when $v = c/\sqrt{2}$, one way of characterizing the speed of light is that it is associated with a relative speed at which the electrical force of mutual attraction/repulsion must vanish. That is a consequence that never gets mentioned in the standard treatments of relativistic electrodynamics.

If one recalls that the roots of the theory of relativity were in the limits of Maxwell's theory of electromagnetism, such as Einstein's discussion of the electrodynamics of moving bodies or Poincaré's theory of the electron, then perhaps it would be fruitful for theoretical physics to revisit those roots in light of Weber's observation.

---

is the speed of light *in vacuo*, he did eventually come to that conclusion.

3.6 Some Personal Notes

Peyman Ghaffari[1,2]
1 - IMAAC-next, Tech Park of Fuerteventura
Puerto del Rosario, Las Palmas 35600, Spain
2 - Center for Research and Development in Mathematics and Applications
Department of Mathematics, University of Aveiro
3810-193 Aveiro, Portugal
E-mail: pgsaid@fc.ul.pt

I am working in the field of mathematical epidemiology trying to understand the dynamics of infectious diseases spreading in a network or society. This also includes application of Optimal Control Theory on deterministic and stochastic epidemiological models describing spreading of mosquito transmitted Vector-borne Diseases using numerical methods. Other scientific interests include: Complex Systems, Self-Organization, Fractional Derivatives and Neuronal Networks.

I received my Ph.D. in Mathematical-Physics (Non-linear Dynamics and Complex Systems) at Imperial College (London) after finishing my M.S. (German Diploma in Physics) in Theoretical Plasma Physics at University of Düsseldorf (Germany). After completing my PhD, I worked as an industrial consultant many years outside academia, which helped me to develop a different perspective towards science.

Being always interested in History of Science, I discovered a paper of Prof. Andre Koch Torres Assis in Internet a few years ago. I was stunned to hear about the Weber Electrodynamics for the first time. Unfortunately, this topic is not taught anymore at universities and is not included in University Curriculum.

Weber's approach is in my opinion of great value (even if the theory shows to be wrong in future experiments) for the understanding of the history of Electrodynamics. I am happy to contribute as translator of Weber's work into English for broader audience helping rediscovering this neglected piece of research.

3.7 Weber and the Manhattan Project: How I Learned about Weber's Electrodynamics

Laurence Hecht
E-mail: larryhecht33@gmail.com

I learned about Ampère's and Weber's electrodynamics from an unusual man who had earned PhDs in both physical chemistry and physics at the University of Chicago in the 1930s. Dr. Robert James Moon (1911-1989) was also an important figure in the Manhattan Project, in part because in 1935 he had built the 50-inch cyclotron, later used in the atomic research, to qualify for his PhD in physical chemistry under William Draper Harkins.[66]

The story of that cyclotron is not well known, but it played an essential role in achieving criticality in the first atomic pile that was built under the football field at the University of Chicago. I learned a lot about this one evening in 1985, when Dr. Moon and Dr. Erich Bagge, who was Werner Heisenberg's assistant on the German effort to achieve a nuclear explosive device, met over my dinner table in Leesburg, Virginia. Like the Americans, the Germans had first sought to use a carbon moderator to absorb the unwanted fast neutrons and leave the slower ones required for fission of the U-235 isotope.

Both the Germans and the Americans had faced the same problem: that to effectively absorb the high-energy neutrons, the carbon blocks must be extremely pure. The German failure to solve it caused them to shift to heavy water as a moderator which required an enormous amount of electrical energy that had to be obtained from Norwegian hydroelectric capacity. That was one of many problems that delayed the German development of the bomb, and Bagge, who it appeared had not shared Heisenberg's reservations about developing the weapon, was very curious to learn how the Americans had done it. Moon, then age 74, was a willing storyteller, while Bagge, the same age, was all ears.

The Chicago branch of the Manhattan project was using carbon blocks produced in the refractory ovens of the main steelworks of the city. Moon suspected that it had to be impurities in those carbon blocks that prevented the moderator from working effectively. Using charged particles produced by his cyclotron he probed the blocks to find how long the neutrons lasted inside, and eventually found that the impurities were concentrated in the outer layers, apparently migrating there during the heating of the blocks. When the blocks were machined down, a very dirty job that left all those involved with the appearance of coalminers, they became effective moderators. Some modifications in the production process at Chicago's South Works steel plant, and careful testing of each block with the cyclotron, then allowed the Chicago pile to go "critical."

The remarkable thing about that cyclotron, as Moon later told me, was that he had designed it using the Weber electrodynamics. He spoke often of the importance of Ampère and Weber's work. Unfortunately he died in 1989 before I was able to undertake a serious study of these topics. Conversations with him often went well beyond my experience and scientific knowledge, but he would patiently answer any question asked of him, and I avidly pursued them up to the very outer edge of my comprehension at the time. All that I

[66] [MH36b] and [MH36a].

recall of any specificity concerning the cyclotron was that he had studied the first cyclotron produced by E. O. Lawrence a year or two earlier and made modifications in the semi-circular electrodes, known as "dees," which greatly increased the power of the accelerator.

Later I learned from a letter-to-the-editor of *Physics Today* (August 1984, p. 73) that Moon had actually proposed the concept of the synchrotron at a University of Chicago seminar in 1939 where the discovery of nuclear fission and the possibilities of a nuclear bomb were first broached by Sam Allison, physicist and later director of the secret MetLab during 1943-44. As Franklin Offner, later of Northwestern University recounted it:[67]

> Moon said to me, "People say that there is a relativistic limit to the power of a cyclotron, due to defocusing with the relativistic increase in mass. But I think it would be easy to overcome this by just frequency modulating the Ds to keep up with the particle mass."

Curiously, Moon's decision to build the cyclotron developed out of a long-term program to achieve thermonuclear fusion that he had worked out with Dr. Harkins not long after arriving at Chicago in 1930.

A few years after Moon's death in 1989, I was able to devote some time to a study of Ampère's and Weber's electrodynamics. My initial study of Weber relied largely upon the *Sixth Memoir* which was one of the few works then available in English.[68] It took several readings and finally a study of the *First Memoir* translated by my friend Susan P. Johnson,[69] before I felt I truly understood what I came to call the Ampère-Gauss-Weber electrodynamics.

But even before I had attained a more complete understanding, several things stood out prominently in the *Sixth Memoir*. Using our modern terminology, Weber had conceptually identified, no later than 1871, the proton-electron mass ratio, the classical electron radius, the nuclear radius, and a plausible configuration for proton or electron pairs. He also discussed a limiting relative velocity, c ($= \sqrt{2} \times$ the speed of light) for two charged particles, the value of which had already been derived in experiments with Kohlrausch more than a decade earlier.[70]

That was more than enough to provoke my abiding interest in Weber, along with a growing sense of injustice over how these profound discoveries had been relegated to the dustbin of scientific history.

3.7.1 Moon's Model of the Nucleus

Moon had first sparked my curiosity about Ampère and Weber's contribution to electrodynamics in 1981, when I collaborated with him in teaching science classes at a youth summer camp. I attempted, with occasional success, to spark some interest in geometry with a class on construction of the Platonic and Archimedean solids. He focused on teaching Ampère's discoveries in electrodynamics, using four simple experiments which he felt best encompassed Ampère's discoveries: the straight wire, parallel wires, twisted wires, and the solenoid.

As it happened, we were there at different times during the summer, so our paths did not cross much. But four years later, when I saw him again, he showed me something

[67] [OM84].
[68] [Web71] with English translation in [Web72].
[69] [Web46] with partial French translation in [Web87] and a complete English translation in [Web07].
[70] [KW57] with English translation in [KW21].

truly remarkable: a model of the atomic nucleus using a nested structure of four of the five Platonic solids to represent the elements to palladium, and a similar, twinned structure to portray those up to uranium. I first wrote this up in an article published in 1988 titled "The Geometric Basis for the Periodicity of the Elements."[71] Over the following decades, as I came to better grasp the Weber electrodynamics, and particularly the concept of a dynamic stable structure of paired charged particles, which I came to call the "Weber pair," I made several new attempts at these geometric constructions.[72]

As Moon later recounted it,[73] the idea had come to him in 1985 while considering the implications of Klaus von Klitzing's experiments demonstrating the quantization of the Hall resistance at extremely low temperature and high magnetic field. Over that summer or early fall Moon had also read Johannes Kepler's extraordinary work, *Mysterium Cosmographicum*, in which a similar structure of nested Platonic solids accounted for the distances of the planets from the Sun. Von Klitzing's work was in the news at the time as he had been awarded the Nobel prize for physics that same year.[74] Somehow, the confluence of these two scientific contributions, separated in time by nearly 400 years, produced this result.

It will take me too far afield to go into the further developments of Moon's nuclear model and how I came to understand the contribution of Weber's electrodynamics to his thinking, long after his death in 1989. Moon left little record of his own work in this regard. That was partly due to an innate modesty, but also I believe to a life-altering experience he once described vividly to me that happened to him in August of 1945.

Moon, along with many of the young scientists involved in the secret research of the Manhattan Project, had been deeply concerned throughout the project, and even before its start, over whether, as he put it, mankind was ready for the unleashing of such enormous force as was contained within the atomic nucleus. He was a signer of the July 17, 1945 petition drafted by Leo Szilard of the Chicago MetLab, pointing out the enormous moral risk associated with atomic weaponry and calling upon President Harry Truman to forego use of the atomic bomb against Japan until the terms to be imposed upon surrender were made public in detail, and Japan were still to refuse to accept them.

Two days after the bombing of Hiroshima, the members of the MetLab team at Chicago were assembled and shown the aerial reconnaissance photographs of what remained of the city. The images produced moral horror in many, some to the point of retching and other physical symptoms. Moon said that he resolved at that point to cease active research in nuclear physics. His subsequent work included the development of the first scanning x-ray microscope, researches in the action potential of the nerve, and studies of the aging brain. He continued teaching physics at Chicago, and in 1958 wrote a proposal (unfunded) for the investigation of the interaction between a steady current and a stationary electric charge using a newly improved quartz fiber torsion microbalance.[75] Yet although he always encouraged me in my curiosity to understand and develop his ideas on atomic structure, I long suspected that his reluctance to publish on the topic stemmed from that August 1945 trauma.

May 6, 2021

---

[71] [Hec88].

[72] [HS04].

[73] [Hec04].

[74] [Moo04b].

[75] [Moo04a].

Chapter 4

Wilhelm Weber's Works Translated into English, French and Portuguese

A. K. T. Assis[76]

I present in this Chapter the main contents of a paper published in 2010 with updated references.[77] The latest versions of some of these translations can be found in the present book on *Wilhelm Weber's Main Works on Electrodynamics Translated into English*. Not all works listed here have been included in the present book. I then considered that it would be relevant to list all of these works here.

Wilhelm Eduard Weber (1804-1891) was one of the main scientists of the XIXth century. His complete works were published in 6 volumes between 1892 and 1894.[78] Here I quote his works and letters known to me which have been translated into English, French and Portuguese.

The joint book of Wilhelm Weber and his brother, the anatomist Eduard Friedrich Weber (1806-1871), originally published in 1836, has recently been translated into English:

- German: Mechanik der menschlichen Gehwerkzeuge. Eine anatomisch-physiologische Untersuchung.[79]

  English: Mechanics of the Human Walking Apparatus.[80]

Between 1836 and 1841 C. F. Gauss (1777-1855) and Weber edited six joint works, published between 1837 and 1843, containing the results of the observations made by the German Magnetic Association, with an addendum.[81] The first joint work has already been translated into English and French:

- German: Resultate aus den Beobachtungen des magnetisches Vereins im Jahre 1836.[82]

[76] Homepage: `www.ifi.unicamp.br/~assis`

[77] [Ass10].

[78] [Web92e], [Web92d], [Web93b], [Web94d], [WW93] and [WW94].

[79] [WW94].

[80] [WW92].

[81] [GW37], [GW38a], [GW39b], [GW40b], [GW41a], [GW43] and [GW40a].

[82] [GW37].

English: Results of the observations made by the Magnetic Association in the year 1836.[83]

French: Sur le magnétisme terrestre et sur l'association pour les observations magnétiques.[84]

Some specific works of Weber translated into English and French can be cited:

1. German: Bemerkungen über die Einrichtung magnetischer Observatorien und Beschreibung der darin aufzustellenden Instrumente.[85]

   French: Remarques sur l'établissement des observatoires magnétiques, et description des instrumens à y placer.[86]

   First English translation: Remarks on the construction of magnetic observatories and the instruments which they should contain.[87]

   Second English translation: Remarks on the arrangement of magnetical observatories, and description of the instruments to be placed in them.[88]

2. German: Beschreibung eines kleinen Apparats zur Messung des Erdmagnetismus nach absolutem Maass für Reisende.[89]

   French: Description d'un petit appareil portatif pour les voyageurs, et destiné aux mesures absolues du magnétisme terrestre.[90]

   English: Description of a small portable apparatus for measuring the absolute intensity of terrestrial magnetism.[91]

3. German: Bermerkungen über die Einrichtung und den Gebrauch des Bifilar-Magnetometers.[92]

   English: Observations on the arrangement and use of the bifilar magnetometer.[93]

4. German: Das transportable Magnetometer.[94]

   English: On a transportable magnetometer.[95]

5. German: Erläuterungen zu den Terminszeichnungen und den Beobachtungszahlen.[96]

   Partially translated into English: An extract from remarks on the term-observations for 1839, of the German Magnetic Association.[97]

[83] [GW41b] and [GW66].
[84] [GW38b].
[85] [Web37a].
[86] [Web38c].
[87] [GW39a].
[88] [Web41f].
[89] [Web37b].
[90] [Web38b].
[91] [Web41a].
[92] [Web38a].
[93] [Web41d] and [Web66b].
[94] [Web39].
[95] [Web41e] and [Web66c].
[96] [Web40].
[97] [Web41b] and [Web66a].

An extremely rich exchange of letters between Gauss and Weber which took place in 1845 has been recently translated into English:[98]

- German: Zur Elektrodynamik.[99]

  English: Text of the Gauss-Weber correspondence.[100]

Weber wrote eight major Memoirs between 1846 and 1878 under the general title *Electrodynamic Measurements* (*Elektrodynamische Maassbestimmungen*).[101] The Eighth Memoir was published only posthumously in his collected works.

All of these eight major Memoirs have already been translated. Here I quote their titles as published in this book, namely:

1. First Memoir in German: Elektrodynamische Maassbestimmungen — Über ein allgemeines Grundgesetz der elektrischen Wirkung.[102]

   English: Electrodynamic Measurements, First Memoir, relating specially to a General Fundamental Law of Electric Action.[103]

   Partially translated into French: Mesures Électrodynamiques.[104]

2. Second Memoir in German: Elektrodynamische Maassbestimmungen insbesondere Widerstandsmessungen.[105]

   English: Electrodynamic Measurements, Second Memoir, relating specially to Measures of Resistance.[106]

3. Third Memoir in German: Elektrodynamische Maassbestimmungen insbesondere über Diamagnetismus.[107]

   English: Electrodynamic Measurements, Third Memoir, relating specially to Diamagnetism.[108]

4. Fourth Memoir in German by R. Kohlrausch and W. Weber: Elektrodynamische Maassbestimmungen insbesondere Zurückführung der Stromintensitäts-Messungen auf mechanisches Maass.[109]

   English: Electrodynamic Measurements, Fourth Memoir, specially Attributing Mechanical Units to Measures of Current Intensity.[110]

---

[98] [Ten96], [Hec96b] and [Hec96a].
[99] [Gau45] and [Gaub].
[100] [GW96].
[101] [Web46], [Web52b], [Web52a], [KW57], [Web64], [Web71], [Web78] and [Web94b].
[102] [Web46].
[103] [Web07].
[104] [Web87].
[105] [Web52b].
[106] [Web21b].
[107] [Web52a].
[108] [Web21c].
[109] [KW57].
[110] [KW21].

5. Fifth Memoir in German: Elektrodynamische Maassbestimmungen insbesondere über elektrische Schwingungen.[111]

   English: Electrodynamic Measurements, Fifth Memoir, relating specially to Electric Oscillations.[112]

6. Sixth Memoir in German: Elektrodynamische Maassbestimmungen insbesondere über das Princip der Erhaltung der Energie.[113]

   English: Electrodynamic Measurements, Sixth Memoir, relating specially to the Principle of the Conservation of Energy.[114]

7. Seventh Memoir in German: Elektrodynamische Maassbestimmungen insbesondere über die Energie der Wechselwirkung.[115]

   English: Electrodynamic Measurements, Seventh Memoir, relating specially to the Energy of Interaction.[116]

8. Eighth Memoir in German: Elektrodynamische Maassbestimmungen insbesondere über den Zusammenhang des elektrischen Grundgesetzes mit dem Gravitationsgesetze.[117]

   English: Electrodynamic Measurements, Eighth Memoir, relating specially to the Connection of the Fundamental Law of Electricity with the Law of Gravitation.[118]

In 1848 Weber published an abridged version of his First major Memoir. This work of 1848 is extremely important as he introduced here for the first time his potential energy which is a function not only of the distance between the interacting electrified particles, but also of their relative radial velocity. This paper has also been translated into English:

- German: Elektrodynamische Maassbestimmungen.[119]

  English: On the measurement of electro-dynamic forces.[120]

Three of his works related specifically to diamagnetism have already been translated into English. One is a paper of 1848:

- German: Über die Erregung und Wirkung des Diamagnetismus nach den Gesetzen inducirter Ströme.[121]

  English: On the excitation and action of diamagnetism according to the laws of induced currents.[122]

---

[111] [Web64].
[112] [Web21a].
[113] [Web71].
[114] [Web72].
[115] [Web78].
[116] [Web21d].
[117] [Web94b].
[118] [Web08].
[119] [Web48a].
[120] [Web52d], [Web66f] and [Web19].
[121] [Web48b].
[122] [Web52c] and [Web66e].

The second one is a paper of 1852, which is an abridged version of Weber's Third major Memoir on Electrodynamic Measurements:

- German: Ueber den Zusammenhang der Lehre vom Diamagnetismus mit der Lehre von dem Magnetismus und der Elektricität.[123]

  English: On the connexion of diamagnetism with magnetism and electricity.[124]

The last one is a letter of September 25, 1855, from Weber to John Tyndall (1820-1893) related to the theory of diamagnetism:

- English: On the theory of diamagnetism — Letter from Professor Weber to Prof. Tyndall.[125]

A paper of 1851 on the measurement of electric resistance according to an absolute standard has been translated into English in 1861:

- German: Messungen galvanischer Leitungswiderstände nach einem absolutem Maasse.[126]

  English: On the measurement of electric resistance according to an absolute standard.[127]

A joint paper by Weber and Rudolf Kohlrausch (1809-1858) of 1856 has recently been translated into English and Portuguese:

- German: Über die Elektricitätsmenge, welche bei galvanischen Strömen durch den Querschnitt der Kette fliesst.[128]

  English: On the amount of electricity which flows through the cross-section of the circuit in galvanic currents.[129]

  Portuguese: Sobre a quantidade de eletricidade que flui através da seção reta do circuito em correntes galvânicas.[130]

Weber's aphorisms published only posthumously have recently been translated into English:[131]

- German: Aphorismen.[132]

  English: Aphorisms.[133]

---

[123] [Web52e].
[124] [Web53] and [Web66d].
[125] [Web55].
[126] [Web51].
[127] [Web61].
[128] [WK56].
[129] [WK03].
[130] [WK08].
[131] [Ten97b].
[132] [Web94a].
[133] [Web97].

Chapter 5

[Gauss, 1832, Abstract of the Paper:] *Intensitas vis magneticae terrestris ad mensuram absolutam revocata*

Carl Friedrich Gauss[134,135]

Feb. 14, 1833. — The following announcement was made from the Chair:

"His Royal Highness the President has received from Professor Gauss the abstract of a paper read by him at the meeting of the Royal Society at Göttingen, on the 15th of December last, entitled '*Intensitas vis magneticae terrestris ad mensuram absolutam revocata.*' Mr. Gauss's views possessing considerable interest, His Royal Highness is desirous that they should be made known to the Fellows of the Royal Society; but as the original paper will not be printed for many months, and the abstract which appeared in the *Göttingische gelehrte Anzeigen* is in a language not generally understood in this country, His Royal Highness has requested your Foreign Secretary to translate it; and I am commanded to desire your Secretary to read the same to the present meeting.

"In deviating thus far from the usual routine of the business of the Royal Society, His Royal Highness is actuated by a wish to promote the reciprocal and early communication of new and important discoveries and views in science, between our own and the other Societies of Europe, devoted, like this, to '*the improvement of natural knowledge*'.

"Communications of this nature, however, cannot of course be admitted into your Transactions; but the publication, from time to time, of your Proceedings, affords a happy means of giving them general circulation; and thus the rapid propagation of much valuable information will be effected, which otherwise, if not absolutely lost to us, would, at least, long remain unknown to the British scientific public."

The following is the abstract of Professor Gauss's Memoir:

134[Gau32] with English translation in [Gau33a] and [Gau37a], see also [Rei02, pp. 138-150].

135The Notes by A. K. T. Assis are represented by [Note by AKTA:].

Of the three elements which determine the manifestation of terrestrial magnetism in a given place, viz. Declination, Inclination, and Intensity, the first soonest engaged the attention of philosophers, the second much later, and the third has only at a very recent period become an object of investigation and experiment. This progressive interest is chiefly to be accounted for by the circumstance, that while the variation of the compass offered the greatest interest, as applied to the purposes of navigation and geodesic operations, the dip was looked upon as more nearly allied to it than was the intensity of terrestrial magnetism. To the natural philosopher, those three elements are absolutely of the same import, inasmuch as our knowledge of the general system of terrestrial magnetism will ever remain imperfect, until an equal share of attention has been bestowed on its separate branches.

For the first light thrown upon this subject we are indebted to the Baron Humboldt, whose attention was particularly directed to it during all his travels, and who has furnished a considerable series of observations, from which the gradual increase of this intensity, from the magnetic equator of the earth towards the magnetic poles, has been deduced.[136] Many observers have since followed the footsteps of that great naturalist; and almost every part of the world to which, in recent times, travellers have penetrated, has furnished its quota of materials, from which already Hansteen (to whom this branch of philosophical inquiry is under great obligation) has been enabled to attempt the construction of an iso-dynamical chart.[137]

The mode adopted in all these observations consists in disturbing the equilibrium of one and the same magnetic needle in places the comparative intensity at which is to be determined, and in exactly measuring the duration of its oscillations. This duration is indeed, *caeteris paribus*,[138] dependent on the magnitude of the arc; but in such a manner, that however small the arc becomes, it still approaches a determined limit, loosely called the duration,[139] and to which, the arc of oscillation being known, the really observed duration may easily be reduced. The intensity of terrestrial magnetism is thus inversely proportional to the square of the duration of oscillation of the same needle,[140] or directly so to the square of the number of oscillations in a given time; and the result relates to the whole force, or to the horizontal portion of it, according as the needle has been caused to vibrate, in the plane of the magnetic meridian, round a horizontal axis, or, in a horizontal plane, round a vertical axis.

It is evident that the admissibility of this method entirely rests on the assumption of the unchanged magnetic state of the needle employed. If a properly-magnetized and carefully-preserved needle of good hardened steel be made use of for the experiments, and these do not take up too long a space of time, the danger to be apprehended from such alteration may not, indeed, be considerable; and the observer may rest the more satisfied in this respect, if, on returning to the first place, he find the time of the vibration to be the same; but experience teaches us that this result cannot by any means be calculated upon; neither can it be denied, that in resorting to such a proof we are only reasoning in a circle. It was known indeed, long ago, that both the declination and inclination in the same place are far from being

[136][Note by AKTA:] Alexander von Humboldt (1769-1859). See: [Hum01], [HB04], [HB05], [Hum29b], [Hum29a], [MB91] and [MKSG10].

[137][Note by AKTA:] Christopher Hansteen (1784-1873). See: [Han25], [Han27] and [RR15].

[138][Note by AKTA:] *Ceteris paribus* is a Latin phrase that is commonly translated as "all else being equal".

[139][Note by AKTA:] In German: *Schwingungsdauer*. This expression can also be translated as "period of oscillation".

[140][Note by AKTA:] That is, inversely proportional to the square of the period of oscillation of the same needle.

invariable; that both of them, in the course of time, undergo very considerable progressive variations, independently of those periodical ones by which the nicety of observation is affected in different seasons and parts of the day. It is, therefore, no matter of doubt that the intensity of terrestrial magnetism must likewise be subject to them; indeed, the periodical diurnal variations are clearly perceptible in delicate observations. Hence, even if, after a considerable lapse of time, the same time of vibration is again observable in a given place, we are not, on that account, warranted in ascribing this circumstance to anything but a casual compensation of the variations which the intensity of the magnetism of the earth in that place, and the magnetic state of the needle itself, may have experienced during that interval. But even allowing the certainty of the comparative method to be only diminished to a certain degree, not entirely annulled, provided too long a space of time do not intervene, that mode, at all events, becomes entirely useless in cases where it is required to ascertain what changes the intensity of terrestrial magnetic force undergoes in a given place during a very long interval. This question, of considerable interest in a scientific point of view, must, therefore, remain unanswered until the merely comparative method shall be superseded by one which reduces the intensity of terrestrial magnetism to unities[141] perfectly determined and manifest, and entirely independent of the individual nature of the needles employed in the experiments.

It is not difficult to lay down the theoretical principles on which such an independent method is to be founded. The time of oscillation of a given needle depends on three quantities; namely, the intensity of the terrestrial magnetism, the static momentum of the free magnetism in the needle, and the momentum of the inertia of this needle. The last of them may readily be ascertained by suitable methods; and thus, from the observed duration of the oscillation, is deduced, not the quantity of the intensity of the terrestrial magnetism, but the product of this quantity into the static momentum of the free magnetism in the needle. But it is impossible to separate these two factors from one another, unless observations of quite a different kind be superadded, that involve a different combination of them; and this end is attained by the use of a second needle, which, in order to ascertain the ratio of these forces, is subjected both to the influence of the magnetism of the earth and to that of the first needle. These two effects do, indeed, partly depend on the magnetic state of the second needle; but, by suitably conducting the experiments, the observer may eliminate that state, inasmuch as the *ratio* of both forces becomes the more independent of it, the greater the distance of the two needles from one another is assumed. Here, however, it is obviously necessary, at the same time, to consider the position relative to the magnetic meridian, of the magnetic axes of both needles, and of that of the straight line connecting their centres, as also the magnetic state of the first needle; all which cannot be subjected to computation unless we know the law of the force exerted on each other by two elements of free magnetism, or, in other words, with which, according as they are of the same or different denominations, they repel or attract each other. Tobias Mayer had already conjectured this law to be the same with that of general gravitation, i.e. that the force is in the inverse ratio of the square of the distance.[142] Coulomb and Hansteen have endeavoured experimentally to confirm this conjecture;[143] and the fact is now completely established by the experiments detailed in

[141][Note by AKTA:] In German: *Einheiten*.

[142][Note by AKTA:] Tobias Mayer (1723-1762). Gauss is referring to Mayer's 1760 work, [May72], [Hei82, pp. 83-85] and [Hei99, pp. 91-93].

[143][Note by AKTA:] Charles Augustin de Coulomb (1736-1806). See: [Cou88] with German translation in [Cou90] and a partial English translation in [Cou35]; [Pot84]; [Gil71b] and [Gil71a]. See also: [Han19].

Professor Gauss's forthcoming memoir. This law, however, only relates to the elementary effect; for the computation of the total effect of a magnetic body on another, as soon as the nature of the distribution of free magnetism in these bodies is accurately known, becomes a problem purely mathematical, and consequently remains dependent on their casual individual nature; but the greater the distance, the less the influence of this individuality becomes; and if the distance be very great, we may, *caeteris paribus*, assume (as indeed follows from the above principle,) the total effect to be inversely proportional to the cube of the distance. The product of this cube into the fraction which expresses the ratio of the effect of the first needle, and of the terrestrial magnetism on the second needle, will therefore, as the distances continually increase, tend to a determined limit. A proper combination of observations at several judiciously selected distances will, being mathematically treated, make us acquainted with that limit, from which may be deduced the *ratio* of those two quantities the product of which was derived from the observed times of vibration. The combination of both results will then obviously give those two quantities themselves.

The experiments for comparing the effects of the magnetism of the earth, and of the first needle on the second, suspended by a thread, may be conducted in two different ways; inasmuch as the latter may be observed either in a state of motion or of rest. The former is best effected by placing the first needle in the magnetic meridian of the second, whereby the time of a vibration of the latter is either increased or diminished, according as poles of the same or of different names are opposed to each other. The comparison of the time of vibration thus changed, with that occasioned by terrestrial magnetism alone, or rather, the comparison of an increased with a diminished one (under opposite directions of the first needle), will then readily lead to the ratio sought. The second mode is that of placing the first needle in such a manner that the direction of its influence on the second makes an angle with the magnetic terrestrial meridian; then the angle of deviation from the meridian, in a state of equilibrium, will equally lead to the knowledge of the ratio sought.[144] And here, too, it is more advantageous to compare with each other two opposite deviations, under opposite positions of the first needle. The most advantageous position of this needle is along a straight line drawn through the middle of the second and perpendicular to the magnetic meridian. The first mode agrees upon the whole with that proposed some years ago by Poisson;[145] but the experiments, as far as we have any record of them, made by some natural philosophers with a view to apply that mode, have either entirely failed, or their results can at best be considered only as imperfect approximations.

Professor Gauss, who has made frequent trials of both those modes of proceeding, is satisfied that the second is, on many accounts, far preferable to the first.

The real difficulty consists in this, that other elements depending on the individual nature of the needles, enter, as well as the value of the limit, into the influences observed. That effect is represented by a series which proceeds by the negative powers of the distance, beginning from the third; where, however, the following terms become more considerable as the distance is smaller. Now those following terms are to be eliminated by means of several observations; but a slight acquaintance with the theory of elimination easily convinces us that unavoidable errors of observation will never fail to endanger the exactness of the results, as the number

[144][Note by AKTA:] In German: *der Ablenkungswinkel von dem Meridian, im Zustande des Gleichgewichts, führt dann gleichfalls zur Kenntniss des verlangten Verhältnisses.* It was translated as: "when the angle of deviation from the meridian, in a state of equilibrium, will equally lead to the knowledge of the ratio sought." Probably the word "when" was a misprint of "then" (*dann* in German).

[145][Note by AKTA:] Siméon Denis Poisson (1781-1840). See: [Poi25a] and [Poi25b].

of coefficients to be eliminated is greater; so that their number need not be very considerable to render the results of computation entirely useless. No precision, therefore, in the results can be expected, unless such considerable distances are employed as will make the series rapidly converge, and a few terms of it suffice. But in this case the effects themselves are too small to be determined with exactness by our present means of observation; and thus the ill success of the experiments hitherto made is readily explained.

However easy, therefore, in theory the methods of reducing the intensity of terrestrial magnetism to absolute unities[146] may appear, yet their application will ever remain precarious until magnetic observations have attained to a much higher degree of precision than they have hitherto possessed. It is with this view that Professor Gauss has followed up several ideas long ago entertained by him relative to the improvement of our means of observing; confidently expecting that magnetic observations will, ere long,[147] be carried to a degree of perfection nearly, if not altogether, equal to that of the most delicate astronomical observations. The expectation has been answered by the result. Two apparatus fitted up in the observatory of Göttingen, and which have been employed for making the observations, of which several are given in his memoir, leave nothing to desire but a suitable locality completely secured from the influence of iron and currents of air.

The following short abstract from the detailed description of the two apparatus and their effect, given in the memoir itself, will no doubt be acceptable to naturalists interested in this kind of research. Professor Gauss has generally employed needles (if prismatic bars of such strength may be designated by that name) of nearly a foot in length, weighing each about one pound. They are suspended by an untwisted thread of $2\frac{1}{2}$ feet in length, composed of thirty-two threads of raw silk, and thus able to carry even double that weight without breaking. The upper end of the thread is tortile, and the degree of torsion is measured by means of a divided circle. To the south or the north end of the needle (according as the locality renders either the one or the other more convenient), a plane mirror is fixed, the surface of which, by means of two adjusting screws, may be placed perpendicular to the axis of the needle; but scrupulous attention need not be paid to this adjustment, as any deviation may most exactly be measured by the observations themselves, and taken into account as errors in collimation. The needle thus balanced is enclosed in a wooden cylindrical box, which, besides the small aperture in the lid for the passage of the thread, has a larger one in the side, which is rather higher and wider than the mirror already mentioned.

Opposite to the mirror, a theodolite is placed, the vertical axis of which is in the same magnetic meridian with the thread of suspension, and at a distance from it of about sixteen Parisian feet. The optical axis of the telescope is placed rather higher than the needle, and inclined in the vertical plane of the magnetic meridian, so as to be directed towards the centre of the mirror on the needle.

To the stand of the theodolite is fixed a horizontal scale of four feet in length, divided into single millimetres: it makes a right angle with the magnetic meridian. That point of the scale which is situated in the same vertical plane with the optical axis of the telescope, and which, for the sake of brevity, may be denominated the zero point, is marked out by a fine thread of gold depending from the middle of the objectglass, and charged with a weight. The scale is fixed at such a height that the image of a portion of it is seen in the mirror through the telescope, the eye-glass of which is adjusted accordingly. At the opposite side from the needle, in the same vertical plane, and at a distance from the telescope equal to

[146][Note by AKTA:] In German: *absolute Einheiten.*

[147][Note by AKTA:] That is, soon or quickly.

that of the image, a mark is fixed, serving every instant to ascertain the unchanged position of the theodolite.

It is obvious, that if all these conditions be fulfilled, the image of the zero point on the scale will appear exactly on the optical axis of the telescope, and that, so far as an object of known azimuth is visible at the place of the theodolite, we may, by means of this instrument, immediately find the absolute magnetic declination. If, on the other hand, those conditions are only partially fulfilled, then, generally speaking, the image, not of the zero point, but that of another point of the scale, will appear on the optical axis; and if the horizontal distance of the scale from the mirror have been measured with exactness, it will be easy to reduce the amount of the divisions of the scale to the corresponding angle, and thus to correct the result first obtained. By turning the needle in the stirrup (so that the upper surface becomes the lower), the amount of the error of collimation of the mirror may be ascertained with great ease and precision. In both the apparatus, one part or division of the scale is equal to nearly twenty-two seconds; an interval which even the least practised eye may easily subdivide into ten parts.

By this mode of operating, therefore, the direction of the needle and its variations are determined with the greatest possible precision. It is by no means necessary always to wait till it is at rest; as the two elongations to the right and the left may be observed with great accuracy, and their combination, properly managed, will indicate the corresponding point of rest with equal precision. During the antemeridional hours, when the daily variation is most rapid, this may be followed almost from one minute of time to the other.

Of equal importance is this mode of proceeding for observing the duration of the vibrations.[148] The passage of the vertical thread in the telescope before a fixed point of the scale (properly speaking, the reverse is the case), may, even if the whole deviation only amount to a few minutes, be observed with such a degree of precision as never to leave any uncertainty amounting to the tenth of a second in time. The considerable duration of a vibration (about 14 seconds in the most intensely magnetized needles), and the slow degrees by which the arc decreases, are productive of other important advantages: only a few vibrations are required to enable us to determine the time of one vibration with such accuracy, that, though the needle be left to itself for one or even several hours, no doubt will remain on the mind of the observer as to the number of oscillations performed during the interval of his absence. We may commence with vibrations so small (such, for instance, as those with which we generally leave off,) that the reduction to infinitely small vibrations becomes almost imperceptible; and yet, after an interval of six and more hours, the vibrations are still found sufficiently great to admit of having their beginnings observed with all requisite precision.

In cases where anomalies still appear in the observations (which, however will prove so trifling, that with the common means they would have been altogether imperceptible,) they are solely to be ascribed to the current of air which, in the locality where the experiments were made, could not be altogether avoided. To remedy this inconvenience the aperture of the box might be closed by a plane glass; but none perfectly true was within the author's reach, neither could it have been made use of without an inconvenient loss of light.

To the enumerated advantages of this method another may be added, which is, that the observer constantly remains at a great distance from the needle, while in the old mode of proceeding his proximity to it was unavoidable; so that, even if enclosed in a glazed case, it was exposed to the disturbing influence which might be exerted upon it by the warmth of

[148] [Note by AKTA:] In German: *für die Beobachtung der Schwingungsdauer.* That is, for the observation of the period of oscillation.

the body, or that of the lamp, by the iron or even the brass which the experimenter might happen to carry about him.

The advantages of stout heavy needles over those of diminutive size, which have been made use of for most magnetical observations, particularly those relating to the time of vibration, are dwelt upon by Professor Gauss; he has since successfully employed one weighing upwards of two pounds, and expresses his conviction, that if needles of from four to six pounds in weight were used, on which slight currents of air would cease to exert any perceptible influence, magnetic observations might attain an exactness and precision unsurpassed by the most delicate astronomical observations. Much stronger threads would indeed be required for suspension, the torsion of which would produce greater reaction; but whatever the strength of the thread may be, the force of torsion[149] must always, and may without any difficulty, be taken into account with the greatest exactness.

The two apparatus may likewise be made use of for another purpose, which, though not immediately connected with the principal subject of the memoir, may still be adverted to in this place. They are the most sensible and convenient galvanometers both for the strongest and weakest energies of the galvanic current. To measure the strongest, it is only required to bring the conducting wire single, and at a considerable distance (at least several feet), into the magnetic meridian below or above the needle; for very weak energies a multiplier is wound round the box containing the needle. Some of the experiments were made with a multiplier of 68 circumvolutions, producing a length of wire equal to 300 feet. No pair of large plates is requisite; a pair of small buttons, or even simply the ends of two different metallic wires dipped in acidulated water, produce a current indicated by the movement of the image along many hundred parts of the scale; but on using a pair of plates of very moderate dimensions, the image of the whole scale, as soon as the circuit is completed, is seen rapidly to dart through the field of vision of the telescope. It is obvious that by this method the measurement of galvanic forces may be conducted with a degree of ease and precision unattainable by the hitherto employed laborious modes by means of observed times of vibration; and it is literally true that by it we are enabled to follow from second to second the gradual decrease of the intensity of a galvanic current, which, it is well known, is more rapid in the beginning. If, in addition, instead of the single, a double (astatic) needle is used,[150] no degree of electro-magnetic energy will be found too small to admit of being still measured with the utmost precision. Here, therefore, a wide field is opened to the naturalist for most interesting investigation.

Not a small portion of this unpublished memoir of Prof. Gauss is taken up by the developement of the mathematical theory; and also by various methods peculiar to the author, such as the determination of the momentum of inertia of the vibrating needle, independently of the assumption of a regular figure; by his experiments with a view to establish the above-mentioned fundamental law for the magnetic effects; and, finally, by the details of the experiments to determine the value of the intensity of terrestrial magnetism, of which last the following may be given as the results, as far as they relate to the intensity of the

[149][Note by AKTA:] In German: *Torsionskraft*.

[150][Note by AKTA:] The adjective "astatic" is used in physics with the meaning of something having no tendency to take a definite position or direction. An astatic needle can be a combination of two parallel magnetized needles having equal magnetic moments, but with their poles turned opposite ways, that is, in antiparallel position. The arrangement protects the system from the influence of terrestrial magnetism. It was invented by Ampère, [Amp21] and [LA98]. An earlier system composed of a single magnetized needle had also been created by Ampère, [Amp20b, p. 198] with Portuguese translation in [CA09, p. 133], [Amp20a, p. 239] and [Amp, p. 2], see also [AC15, p. 57].

horizontal part of that force.

I.	May 21	1.7820
II.	May 24	1.7694
III.	June 4	1.7713
IV.	June 24-28	1.7625
V.	July 23, 24	1.7826
VI.	July 25, 26	1.7845
VII.	Sept. 9	1.7764
VIII.	Sept. 18	1.7821
IX.	Sept. 27	1.7965
X.	Oct. 15	1.7860

For unities, the millimetre, the milligramme, and the second in time have been adopted. The manner in which the measurement of the intensity has been determined by them cannot here be specified: the numbers, however, remain the same, provided the unity of space, and that of weight (properly speaking, unity of masses), are changed in the same proportion. These experiments vary partly with regard to the greater or less degree of care with which they were conducted, partly with regard to the places in which they were made, and to the needles employed.

The experiments VII, VIII, IX, were in every respect performed with all the precision which the apparatus in the present state admits of, and the distances were measured with microscopic exactness. In experiments IV, V, VI, X, some operations have been performed with rather less care; and the first three experiments are still less perfect in this respect.

The needles employed in the first eight experiments were not indeed the same, but they were nearly alike in size and weight (the latter between 400 and 440 grammes); the principal needle in experiment X weighs 1062 grammes; experiment IX on the other hand was made, with a much smaller needle (weight 55 grammes), merely for the sake of ascertaining the degree of precision, which, all other precautionary means being alike, may be attained in using a needle of such small dimensions: the result of this experiment is therefore much less to be depended upon.

Experiments VII to X were made in one and the same place in the observatory; the preceding ones in other places in the same observatory, and in apartments of the author's dwelling-house. No perfectly pure results therefore could be derived from these latter experiments, inasmuch as the iron in those localities, and particularly in the observatory, becoming itself magnetic by the magnetism of the earth, would necessarily react upon the needle, and confound its influence with that of the terrestrial magnetism. Such places, indeed, were uniformly chosen in which neither fixed nor moveable masses of iron were near; nevertheless, even the more distant ones may not have been altogether without their effect upon the operations. However, on casting a look over the different results, it appears probable, that in no one of those localities, the modification of the terrestrial magnetism produced by extraneous influence exceeds the hundredth part of the whole. But results commensurate to the precision belonging to this mode can only be expected in a locality entirely free from the influence of iron.

In order to obtain the intensity of the *whole* force of the terrestrial magnetism, the numbers found are to be multiplied by the secant of the inclination. Mr. Gauss intends at a future period also to treat this element according to peculiar methods; in the mean time he merely mentions that on June the 23rd he has found $68^\circ\ 22'\ 52''$ with the inclinatorium of the University collection of instruments, — a result which, as the observation was made in the observatory, and therefore not without the reach of local interference, may possibly require to be rectified by other observations.

Chapter 6

Editor's Introduction to Gauss' Work on the Absolute Measure of the Earth's Magnetic Force

A. K. T. Assis[151]

Gauss' work on the absolute measure of the Earth's magnetic force was announced with an abstract of the full paper at the Königlichen Societät der Wissenschaften zu Göttingen in December 1832, see Chapter 5.

The original paper in Latin was published only in 1841, although a preprint appeared already in 1833 in small edition.[152] Several translations have been published. There are two German versions, one translated by J. C. Poggendorff published in 1833, the second one translated by A. Kiel with notes by E. Dorn and published in 1894; a French version by Arago in 1834; two Russian versions, one by A. N. Drašusov of 1836 and another one by A. N. Krylov in 1952; an Italian version by P. Frisiani in 1837; an English extract was published in 1935, while a complete English translation by S. P. Johnson (1995) was published in 2003; and a Portuguese version translated by A. K. T. Assis in 2003.[153]

Although the original paper was published only in 1841, I am considering here that it became known in 1833 since a preprint circulated in small edition and a full translation into German was published in 1833.

The title of the English extract published in 1935 was "The absolute measure of magnetic force".[154] The title of the English version of 1995 was "The intensity of the Earth's magnetic force reduced to absolute measurement".[155] The title of the English version appearing in this book has been chosen as "The absolute measure of the Earth's magnetic force".

The English version presented in this book is based on the translation from the German by the late Susan P. Johnson (1995).[156] It was edited by Laurence Hecht and A. K. T. Assis.

[151]Homepage: www.ifi.unicamp.br/~assis
[152][Gau41a] and [Rei19].
[153][Gau33b], [Gau34c], [Gau36a], [Gau37c], [Gau94], [Gau35], [Gau52], [Gau75], [Gau03] and [Ass03b].
[154][Gau35].
[155][Gau03].
[156][Gau94] with English translation in [Gau03], see also [Joh97].

Chapter 7

[Gauss, 1833] The Absolute Measure of the Earth's Magnetic Force

Carl Friedrich Gauss[157,158]

[The treatise "Intensitas vis magneticae terrestris ad mensuram absolutam revocata" was read by *Gauss* at the Göttingen Gesellschaft der Wissenschaften (Royal Scientific Society) on December 15, 1832, and printed in Volume 8 of the Treatises of this Society, pp. 3-44.[159]

The translation from the Latin to the German was provided by Herr Oberlehrer Dr. *Kiel* in Bonn. Edited by E. Dorn, Leipzig, Wilhelm Engelmann Verlag, 1894.][160]

For the complete determination of the Earth's magnetic force at a given location, three elements are necessary: the deviation (declination) or the angle between the plane, in which it acts, and the meridian plane; the inclination of the direction to the horizontal plane; finally, third, the strength (intensity). The declination, which is to be considered as the most important element in all applications to navigation and geodesy, has engaged astronomers and physicists from the beginning, who for a century, however, have given their constant attention to the inclination as well. In contrast, the third element, the intensity of the terrestrial magnetic force, which is surely just as worthy a subject of science, remained fully neglected until more recent times. *Humboldt* rendered the service, among so many others, of having been first to direct his attention to this subject, and in his travels assembled a large array of determinations concerning the relative strength of terrestrial magnetism, which yielded the result of a continuous increase in this strength from the magnetic equator toward the pole.[161] A great many physicists have trod in the footsteps of that great natural scientist,

[157] The English version presented in this book is based on the translation from the German text, [Gau94], by the late Susan P. Johnson (1995), [Gau03], see also [Joh97]. It was edited by Laurence Hecht and A. K. T. Assis.

[158] Carl Friedrich Gauss' Notes are represented by [Note by CFG:]; the Notes by E. Dorn, the Editor of Gauss' work published in German in 1894, are represented by [Note by ED:]; while the Notes by A. K. T. Assis are represented by [Note by AKTA:].

[159] [Note by AKTA:] [Gau41a].

[160] [Note by AKTA:] This Note appeared on the 1894 publication of the German translation of this work, [Gau94, p. 49].

[161] [Note by AKTA:] See footnote 136 on page 38.

and have already brought together such a large array of determinations, that *Hansteen*, highly distinguished for his knowledge of terrestrial magnetism, has recently been able to publish a comprehensive isodynamic map.[162]

The method applied in all these investigations consists of observing either the length of time it takes for one and the same magnetized needle to perform the same number of oscillations at different locations, or the number of oscillations of the same needle within the same time period, and the strength is assumed to be proportional to the square of the number of oscillations in a given length of time: in this way all the intensities are compared with one another, when an inclination-needle, suspended at the center of gravity, oscillates on a horizontal axis perpendicular to the magnetic meridian, or the horizontal components, when a horizontal needle swings on a vertical axis. The latter mode of observation leads to greater precision, and the results emerging from it can, after ascertaining the inclination, easily be related to the total intensities.

It is evident that the reliability of this procedure depends on the assumption, that the distribution of free magnetism in the particles of the needle used in this comparison remains unchanged during the individual experiments; that is, if the magnetic force of the needle had undergone any sort of weakening in the course of time, it would subsequently oscillate more slowly, and the observer, who has no knowledge of such an alteration, would attribute too low a value to the strength of the terrestrial magnetism for the subsequent location. If the experiments span only a moderate period of time and a needle manufactured of well-tempered steel and carefully magnetized is used, a significant weakening of the force is not especially to be feared; moreover, the uncertainty will be still further decreased, if several needles are employed for comparison; finally, this assumption will be relied upon more confidently, if it is found, after returning to the first location, that the duration of the needle's oscillation has not changed. Yet whatever precautions may be taken, a slow weakening of the force of the needle can scarcely be prevented, and thus such conformity after a longer absence can seldom be expected. Therefore, in comparing intensities for widely distant locations on the Earth, precision of the degree we must desire cannot be attained.

After all, this disadvantage in the method is less consequential, so long as it is only a matter of comparison of simultaneous intensities or intensities corresponding to time periods not remote from one another. But because experience has taught us, that both the declination and inclination undergo continuous changes at a given location, it cannot be doubted, that the intensity of terrestrial magnetism is subject to analogous secular changes, as it were. It is evident that, as soon as this question arises, the method described above loses all utility. And yet it would be extraordinarily desirable for the progress of science, that this highly important question be fully settled, which certainly cannot occur, if another method does not replace that purely comparative one, which is independent from the random inequalities of the needles, and the intensity of the Earth's magnetism is reduced to fixed units and independent measure.[163]

It is not difficult, to specify the fundamental theoretical principles, on which such a long-desired method must be based. The number of oscillations, which a needle carries out in a given length of time, depends on the intensity of the Earth's magnetism, as well as the state of the needle, namely on the static moments[164] of the elements of free magnetism contained in it and on its moment of inertia. Because this moment of inertia can be ascertained

[162][Note by AKTA:] See footnote 137 on page 38.
[163][Note by AKTA:] In German: *auf feststehende Einheiten und unabhängige Maasse zurückführt.*
[164][Note by AKTA:] In German: *von dem statischen Momente.*

without difficulty, it is evident that observation of the oscillations would provide us with the product of the intensity of the Earth's magnetism into the static moment of the needle's magnetism: but these two magnitudes cannot be separated, if observations of another kind are not made in addition, which yield a different relationship between them. This goal can be attained, if a second needle is added as an auxiliary and is exposed to the influence of the Earth's magnetism and of the first needle, in order to ascertain the relation of these two forces to each other. Each of the two effects will of course depend on the distribution of the free magnetism in the second needle: but the second will further depend on the state of the first needle, the distance between the midpoints, the position of the straight lines connecting their midpoints, and finally on the laws of magnetic attraction and repulsion. *Tobias Mayer* had been the first to advance the suggestion,[165] that this law accords with the law of gravitation, insofar as those forces also decrease by the square of the distance: the experiments by *Coulomb* and *Hansteen* have given this suggestion great plausibility,[166] and the latest experiments elevate it beyond any doubt. But it is well to consider, that this law refers only to the individual elements of free magnetism; the total force of a magnetic body will be completely different, and, given very great distances, as can be deduced from just that law, will be very nearly proportional to the inverse relationship of the cube of the distance, so that, other conditions being equal, the action of the needle multiplied by the cube of the distance, given ever-increasing distances, approaches a definite limit. This limiting value, as soon as a definite length is taken as the unit, and the distances are expressed numerically, will be of the same type as the action of the Earth's force[167] and comparable with it.

By means of experiments appropriately constructed and performed, the limiting value of this relationship can be determined. Since the limit contains only the static moment of the magnetism of the first needle, then the quotient of this moment divided by the intensity of the terrestrial magnetism will be obtained; if they are now compared with the already ascertained product of these magnitudes, it will serve to eliminate this static moment, and will yield the value of the intensity of the terrestrial magnetism.

With regard to the possible ways to test the actions of terrestrial magnetism and of the first needle on the second needle, a twofold path is open, since the second needle can be observed either in a state of motion or in a state of equilibrium. The first method consists of observing this second needle's oscillations, while the action of the terrestrial magnetism is associated with the action of the first needle. This first needle must be set up at a suitable distance such that its axis lies in the magnetic meridian going through the midpoint of the oscillating needle: thereby the oscillations are either accelerated or retarded, according to whether unlike or like poles are turned toward each other, and the comparison of either the oscillation times for each of the two positions of the first needle with each other, or the oscillation time of one of the two positions with the oscillation time which (after the distancing of the first needle) takes place under the exclusive action of the Earth's magnetism, will show us the relation of this force to the action of the first needle. In the second method, the first needle is placed so that the direction of its force, which it exerts on the location of the second, freely suspended, needle, forms an angle (for example, a right angle) with the magnetic meridian; by this means the second needle itself will be deflected out of the magnetic meridian, and from the magnitude of the deviation, one can infer the relation between the terrestrial magnetic force and the influence of the first needle.

[165][Note by AKTA:] See footnote 142 on page 39.
[166][Note by AKTA:] See footnote 143 on page 39.
[167][Note by AKTA:] That is, the Earth's gravitational force.

By the way, the first method essentially coincides with that proposed some years ago by *Poisson*.[168] But the experiments carried out by certain physicists according to this formulation, at least so far as I know, either totally miscarried, or can at most be considered imperfect approximations.

The actual difficulty lies in the fact that from the observed influences of the needle at moderate distances, a limit must be calculated, which bases itself on, in a sense, an infinitely great distance, and that the eliminations necessary for this purpose are all the more disturbed by the smallest errors in observation, indeed rendered totally useless, the more that unknown [quantities], which depend on the specific condition of the needle, must be eliminated: the calculation can only be reduced to a small number of unknowns, however, when the influences occur at distances, which in relation to the length of the needle become rather large, and therefore they themselves become very small. Yet, in order to measure such small influences the practical expedients employed up to now are insufficient.

I recognized, that I had to turn my efforts above all toward discovering new expedients, whereby the alignments of the needle could be observed and measured with far greater precision than before. The labors undertaken to this purpose, which were pursued for several months, and in which I was assisted by *Weber* in many ways,[169] have led to the desired goal, to the extent that they not only did not disappoint expectations, but far exceeded them; and that nothing more remains to be desired, in order to make the precision of the experiments equivalent to the acuteness of astronomical observations, but a site fully protected from the influence of nearby iron and air currents. Two pieces of equipment were put at our disposal, which are distinguished no less by their simplicity than by the precision they afford. I must reserve their description for another time, while I submit to the physicists in the following treatise the experiments carried out to date in our observatory toward determining terrestrial magnetic intensity.[170]

7.1

To explain magnetic phenomena, we assume two magnetic fluids: one we call north, the other south. We presuppose, that the elements of the one fluid attract those of the other, and that on the other hand, two elements of the same fluid mutually repel each other, and that each of the two forces alters in inverse relation to the square of the distance. It will be shown below that the correctness of this law was itself confirmed by our observations.

These fluids do not occur independently, but only in association with the ponderable particles of such bodies which take on magnetism, and their actions express themselves either when they put the bodies into motion or they prevent or transform the motion, which other forces acting on these bodies, e.g. the force of gravity, would elicit.

Hence the action of a given amount of magnetic fluid on a given amount of either the same or the opposite fluid at a given distance is comparable to a given motive force, i.e. with the action of a given accelerating force on a given mass, and since the magnetic fluids themselves can be known only through the effects, which they bring forth, the latter must directly serve to measure the former.

---

[168][Note by AKTA:] See footnote 145 on page 40.

[169][Note by AKTA:] Wilhelm Eduard Weber (1804-1891).

[170][Note by AKTA:] A general discussion of this method can be found in Chapter 59 (Measurement of horizontal intensity of the Earth's magnetism by Gauss's method) of Friedrich Kohlrausch's book, [Koh83].

In order, however, that we may be able to reduce this measurement to definite concepts, units must above all be established for three kinds of magnitudes, namely, the unit of distance, the unit of ponderable mass, and the unit of accelerating forces. For the third, the force of gravity at the locus of observation can be assumed: if, however, this is not suitable, the unit of time must also enter in, and for us that accelerating force will be $= 1$, which, within the time unit, produces a change of velocity of the body in the direction of its motion, which is equivalent to the unit.

Correspondingly, the unit of the amount of north fluid will be that whose repulsive force on another like it, and whose existing amount of motive force in the unit of distance $= 1$, i.e. the action of an accelerating force $= 1$ on a mass $= 1$; the same will be true of a unit of the amount of south fluid; in this definition, clearly the active fluid, as well as that which is being acted on, must be thought of as, at bottom, concentrated in physical points. Beyond this, however, it must be assumed, that the attraction between given quantities of different kinds of fluids at a given distance is equal to the repulsion between the same respective quantities of the same kind of fluid. Hence the effect of a quantity m of north magnetic fluid on a quantity m' of the same fluid at distance r (each of the two fluids being assumed to be concentrated as at one point) will be expressed as $\frac{mm'}{rr}$, or it is equivalent to a motive force $= \frac{mm'}{rr}$, which acts in the direction of the first against the second fluid, and evidently this formula holds true in general, when, as from now on we wish to stipulate, a quantity of southern fluid will be considered as negative, and a negative value of the force $\frac{mm'}{rr}$ will signify attraction.

Hence if equal quantities of north and south fluid are found simultaneously at one physical point, no action at all will arise; if, however, the amounts are unequal, only the excess of the one which we wish to term free magnetism (positive or negative) will come under consideration.

7.2

To these fundamental presumptions we must add still another, which is confirmed at all times by experience, namely, that every body, in which magnetic fluid occurs, always contains an equal amount of each of the two. Experience even shows that this assumption is to be extended to the smallest particle of such a body, which can still be differentiated by our senses. Yet, according to what we have emphasized at the end of the preceding Section, an action can only be present insofar as some separation of the fluids takes place, so that we must necessarily assume, that this occurs through such small interstices, that they are inaccessible to our measurements.

A magnetizable body must thus be conceived of as the union of innumerable particles, of which each contains a certain quantity of north magnetic fluid and an equally large amount of south, specifically, so that they are either homogeneously mixed together (the magnetism is latent), or have undergone a lesser or greater separation (the magnetism is developed), a separation, however, which can never involve an overflow of fluid from one particle to another. It makes no difference, whether one assumes, that a greater separation originates from a greater amount of freed-up fluid or from a greater gap between them: it is evident, though, that, in addition to the size of the separation, its directionality must come into consideration at the same time, for it is according to whether or not this is in conformity in the different particles of the body, that a greater or lesser total action can arise respecting the points outside the body.

Yet, however the distribution of free magnetism within the body may behave, one can always insert in its place, as a result of a general theorem, according to a specific law, a different distribution on the surface of the body, which exerts fully the same outward force as the first distribution, so that an element of magnetic fluid situated anywhere on the outside experiences exactly the same attraction or repulsion from the actual distribution of magnetism inside the body, as from the distribution thought of as on its surface.[171,172] The same fiction can be extended to two bodies, which, according to the proportion of the free magnetism developed in them, act upon each other, so that for each of them the distribution thought of as on the surface can replace the actual internal distribution. In this way we can finally give its true meaning to the common mode of speech, which, e.g., ascribes exclusively north magnetism to the one end of a magnetized needle, and south magnetism to the other, since it is evident that this turn of phrase does not contradict the principle enunciated above, which is unconditionally demanded by other phenomena. But it may suffice to have noted this in passing; we will discuss the principle itself more fully on another occasion, since it is not required for present purposes.

7.3

The magnetic state of a body consists in the relation of the distribution of free magnetism in its individual particles. With regard to the variability of this state, we perceive an essential difference between the different magnetizable bodies. In some, e.g. in soft iron, that state changes immediately as a result of the slightest force, and if this force ceases, returns to its previous state: in contrast, in others, especially tempered steel, the force must have attained a certain strength, before it can elicit a perceptible change in the magnetic state, and if the force ceases, either the body remains in the transformed state, or at least it does not fully return to the previous one. Hence, in bodies of the first kind, the magnetic molecules order themselves in such a way that, between the magnetic forces, which originate partly from the bodies themselves, partly from external sources, a complete equilibrium exists, or at least the state differentiates itself barely noticeably from that just described. In bodies of the second kind, in contrast, the magnetic state can be lasting even without complete equilibrium between those forces, provided that stronger external forces remain distant. Even if the source of this phenomenon is unknown, it can nevertheless be represented, as if the ponderable parts of a body of the second kind counterpose to the movement of the magnetic fluids associated with it an obstacle akin to friction, a resistance, which in soft iron either does not exist at all or is only very slight.

In theoretical investigation, these two cases require a completely different treatment; but in the present treatise, only bodies of the second kind will be spoken of: in the experiments we want to talk about, the unchangeability of the state in the individual bodies will be a fundamental assumption, and hence one must be on guard, during the experiments to bring other bodies, which could alter this state, into too close proximity.

Yet a certain source of change is at hand, to which the bodies of the second kind are also subjected, namely, heat. Experience indubitably teaches us that the magnetic state of a

[171][Note by ED:] Cf. Gauss, General Principles..., Section 36. *Works*, vol. V, p. 240 (also Ostwald's Classics, No. 2, p. 49).

[172][Note by AKTA:] [Gau40, p. 240 of Vol. V of Gauss' *Werke*] with English translation in [Gau43, p. 195].

body changes with its temperature, yet in such a way that, if the body is not immoderately heated, it returns to its former magnetic state when its former temperature returns. This function is to be determined by means of suitable experiments, and if observations at different temperatures are undertaken as part of an experiment, they will be above all reducible to one and the same temperature.

7.4

Independently of the magnetic forces, which we see sufficiently proximate individual bodies exert upon each other, another force acts upon the magnetic fluids, which, since it manifests itself everywhere on Earth, we ascribe to the globe itself, and hence we term it terrestrial magnetism. This force expresses itself in a dual way: bodies of the second kind, in which magnetism is excited, are, if supported at the center of gravity, turned toward a definite direction: in bodies of the first kind, in contrast, the magnetic fluids are automatically separated by that force, a separation which can be made very noticeable, if one chooses bodies of appropriate shape and puts them in an appropriate location. Each of the two phenomena will be explained by the conception, that at any arbitrary locus, that force drives the north magnetic fluid in a definite direction, and drives the south magnetic fluid, on the other hand, with equal strength in the opposite direction. The direction of the former is always understood, if we speak of the direction of terrestrial magnetism; hence it will be determined both by the inclination to the horizontal plane as by the deviation of the vertical plane, in which it acts, from the fixed meridian plane; the later is called the magnetic meridian plane. The intensity of terrestrial magnetism, however, is to be measured by the motive force which it exerts on the unit of the free magnetic fluid.

This force is not only different at different spots on the Earth, but also changeable at the same spot, in the course of centuries and years, as well as in the course of decades and hours. This changeability has indeed long been known with regard to direction; but with regard to intensity, until now it has only been possible to observe it in the course of the hours of one day, since we had no experimental device designed for longer periods of time. The reduction of the intensity to absolute measure[173] will in future provide remedies for this deficiency.

7.5

In order to subject the action of terrestrial magnetism on magnetic bodies of the second kind (from now on, those are always the kind to consider) to calculation, such a body is conceived of as divided into infinitely small parts; let dm be the element of free magnetism in a particle, whose coordinates in relation to three fixed planes, perpendicular to each other, in the body, may be designated x, y, z: the elements of the south fluid we take to be negative. Then it is clear, first of all, that the integral taken for the entire body (even for every measurable part of the body), is $\int dm = 0$. We wish to further specify $\int xdm = X$, $\int ydm = Y$ and $\int zdm = Z$, which magnitudes may designate the moments of free magnetism[174] in relation to the three orthogonal-planes or in relation to the axes perpendicular to them. Since under the assumption, that a refers to an arbitrary constant magnitude, $\int(x-a)dm$ will be $= X$, it is evident, that the moment in relation to a given axis depends only on its direction, but

[173] [Note by AKTA:] In German: *auf absolutes Maass.*

[174] [Note by AKTA:] In German: *die Momente des freien Magnetismus.*

not on its origin. If we draw a fourth axis through the origin of the coordinates, which forms with the first [three axes] the angles A, B, C, the moment of the elements dm in relation to this axis will be $= (x\cos A + y\cos B + z\cos C)dm$, and further the moment of free magnetism in the body as a whole

$$= X\cos A + Y\cos B + Z\cos C = V \ .$$

If we assume

$$\sqrt{(XX + YY + ZZ)} = M$$

and

$$X = M\cos\alpha \ ; \quad Y = M\cos\beta \ ; \quad Z = M\cos\gamma \ ,$$

and draw a fifth axis, which forms the angles α, β, γ with the first three axes, and with the fourth [axis] the angle ω, then, since as a consequence of these assumptions

$$\cos\omega = \cos A\cos\alpha + \cos B\cos\beta + \cos C\cos\gamma \ ,$$

V will be $= M\cos\omega$. We call this fifth axis simply the *magnetic axis* of the body, and we assume that its *direction* is part of the positive value of the root $\sqrt{(XX + YY + ZZ)}$. If the fourth axis coincides with this magnetic axis, the moment becomes $V = M$, which is clearly the greatest of all moments: the moment in relation to an arbitrary other axis is found, by multiplying this greatest moment (which, so that ambiguity need not be feared, can simply be called the moment of magnetism)[175] with the cosine of the angle between it and the magnetic axis. The moment in relation to an arbitrary axis perpendicular to the magnetic axis will $= 0$, but will be negative in relation to that axis which forms an obtuse angle with the magnetic axis.

7.6

If a force of constant intensity and direction acts on the individual particles of the magnetic fluid, the total resultant force on the body can easily be inferred from the principles of statics, since in the bodies under consideration these particles have lost their state of fluid to some extent, and form a fixed mass with the ponderable body. On an arbitrary magnetic particle dm the motive force $= Pdm$ may act in a direction D (where for the molecule of south fluid the negative sign as such signifies the opposite direction); let A and B be two points on the body lying in the direction of the magnetic axis, and their distance $= r$, positively taken, when the magnetic axis is directed from A toward B: then one easily sees, that if two new forces are added to these forces, each $= \frac{PM}{r}$, of which the one acts on A in direction D, the other on B in the opposite direction, there will be an equilibrium among all these forces. Therefore the former forces will be equivalent to two forces $= \frac{PM}{r}$, of which one acts on B in direction D, the other on A in the opposite direction, and clearly these two forces cannot be united into one.

If, in addition to force P, another similar force P' acts in direction D' on the magnetic fluids of the body, then they can be replaced by two others, which act either on the same

[175] [Note by AKTA:] In German: *das Moment des Magnetismus*. This expression can also be translated as the magnetic moment.

points A and B or more generally on other points A' and B', provided only that $A'B'$ is likewise a magnetic axis, and, if the distance is made $A'B' = r'$, these forces must of course $= \frac{P'M}{r'}$, and must act on B' in direction D', on A' in the opposite direction. The same holds true of several forces.

To the terrestrial magnetic force can safely be ascribed, within a such a small space as that occupied by the body subjected to the experiment, an intensity and direction that are everywhere constant, though variable with time; thus, what we have just said can be applied to it [the terrestrial magnetic force]. However, it can be advantageous, to separate it right at the outset into two forces, a horizontal $= T$ and a vertical $= T'$, which in our region is directed downward. Since, in case one wants to use for the latter two others acting on points A' and B', it is permissible to arbitrarily posit point A' and also the interval $A'B' = r'$, we want to choose A' for the center of gravity, and, denoting by p the weight of the body, i.e. the motive force of gravity on its mass, we say that $\frac{T'M}{p} = r'$. Hereby the effect of force T' is released into a force $= p$ directed upward on A', and into another of the same magnitude directed downward on B', and since further the former is clearly cancelled by the force of gravity itself, the effect of the vertical component is simply reduced to the transfer of the center of gravity from A' to B'. Moreover, it is clear, that for those regions, where the terrestrial magnetic force forms an acute angle with the vertical force, or, in other words, where its vertical part pushes the magnetic north-fluid upward, a similar shift of the center of gravity occurs from the magnetic axis to the south pole.

In this way of thinking, it is self-evident that, whatever sort of experiments may be conducted with a magnetized needle in a single magnetic state, it is impossible to infer the inclination from this alone, but the locus of the actual center of gravity must already be known by some other means. This locus is ordinarily determined before the needle is magnetized; but this practice is not reliable enough, since as a rule a steel needle already takes on magnetism, though weakly, while it is being manufactured. It is therefore necessary for the determination of inclination to induce another shift of the center of gravity by means of an appropriate alteration in the magnetic state of the needle. In order that this differ as much as possible from the first, it will be necessary to reverse the poles, by means of which a double shift is obtained. However, shifting the center of gravity even in needles which have the most suitable form and are saturated with magnetism, cannot exceed a certain limit, which equals roughly 0.4 *mm* in our region, and in regions, where the vertical force is at its greatest, it remains below 0.6 *mm*; from that you can see at the same time, an as great mechanical fineness is required in the needle that is to serve to determine the inclination.

7.7

When any point C of a magnetic body is assumed to be fixed, the necessary and sufficient condition for equilibrium, is that a plane laid through C, the center of gravity and the magnetic axis coincides with the magnetic meridian, and that, moreover, the moments, with which the terrestrial magnetic force and the center of gravity seek to rotate around point C, cancel each other: the second condition proceeds from the fact that, if T denotes the horizontal part of the terrestrial magnetic force and i the inclination of the magnetic axis toward the horizontal plane, $TM \sin i$ is the product of the weight of the body at the distance of the displaced center of gravity B' of the vertical line drawn through C: clearly this distance must lie on the north or south side, according to whether i is an elevation or depression, and for $i = 0$, B' lies in this vertical line itself. If the body is already moved around this vertical line so that the magnetic axis reaches into the vertical plane, whose magnetic azimuth, i.e. its angle with the northerly part of the magnetic meridian (arbitrarily taken as positive toward east or toward west) $= u$, then the terrestrial magnetism will exert a force on the body revolving around the vertical axis, which strives to decrease the angle u and whose moment will be $= TM \cos i \sin u$, and the body will perform oscillations around this axis, whose duration can be calculated according to known methods. Namely, if K denotes the moment of inertia of the body in relation to the axis of oscillation (i.e., the sum of the ponderable molecules multiplied by the square of the distance from the axis), and, as usual, denotes as π the half-circumference for radius $= 1$, then the duration of an infinitely small oscillation will be[176]

$$= \pi \sqrt{\frac{K}{TM \cos i}} ,$$

namely, in the case that it is based upon the magnitudes T and M as unit of the accelerating force, which in the time-unit produces the velocity $= 1$: the reduction of finite oscillations to infinitely small ones will be calculable in a similar way for the oscillations of the pendulum. Hence, if the duration of *one* infinitely small oscillation is found from observation to $= t$, we will have the equation:

$$TM = \frac{\pi\pi K}{tt \cos i} ,$$

and when, moreover, as we always assume from now on, the body is suspended in such a way that the magnetic axis is horizontal:

$$TM = \frac{\pi\pi K}{tt} .$$

If one would rather assume gravity as unit of the accelerating forces, one must divide that value by $\pi\pi l$, where l denotes the length of the simple seconds pendulum, so that altogether one would have:

$$TM = \frac{K}{ttl \cos i}$$

or for our case

$$TM = \frac{K}{ttl} .$$

7.8

When these sorts of experiment are performed on magnetized needles, suspended by a vertical thread, then the reaction, which is exerted by the torsion, cannot be neglected in more refined experiments. We want to identify two horizontal diameters in such a thread, the one, D, at the lower end, where the needle is attached, parallel to the magnetic axis of the needle, the other, E, at the upper end, where the thread is secured, and in fact E would be parallel with

[176][Note by AKTA:] Gauss and Weber utilized the old French definition of the period of oscillation t which is half of the English definition of the period of oscillation T, that is, $t = T/2$, [Gil71a, pp. 154 and 180]. For instance, the period of oscillation for small oscillations of a simple pendulum of length ℓ is $T = 2\pi\sqrt{\ell/g}$, where g is the local free fall acceleration due to the gravity of the Earth, while $t = T/2 = \pi\sqrt{\ell/g}$.

D if the thread were not twisted. We want to assume, that E forms the angle v with the magnetic meridian, while D, or the magnetic axis, forms the angle u with the meridian; then, experience shows that the torsional force[177] will be proportional to the angle $v - u$, at least approximately: hence we will posit the moment, with which this force seeks to make angle u equal to angle v, as $= (v - u)\Theta$.[178] Since the moment of terrestrial magnetic force, which strives to decrease angle u, is now $= TM \sin u$, the condition for equilibrium is contained in the equation:

$$(v - u)\Theta = TM \sin u \ ,$$

which allows for more real solutions, the smaller Θ is in comparison with TM: however, so long as only small values of u are dealt with, instead of this equation one can safely take the following one:

$$(v - u)\Theta = TMu$$

or

$$\frac{v}{u} = \frac{TM}{\Theta} + 1 \ .$$

In our apparatus, the upper end of the thread is fastened to a movable arm, which holds an indicator playing on the periphery of a circle divided into degrees. Hence even if the collimation error (i.e., the point corresponding to the value $v = 0$) is still not known with sufficient precision, nevertheless this pointer indicates the difference for each second value from v on: just as another part of the apparatus provides the difference between the values of u, corresponding to the equilibrium condition, with the greatest precision, and it is clear, that the value of $\frac{TM}{\Theta} + 1$ will be obtained from the division of the difference between two values of v by the difference between the corresponding values of u. If there is a somewhat longer period of time between the experiments conducted for this purpose, it will be necessary, in order to achieve the utmost precision, to take note of the daily change in the magnetic declination, which is easily accomplished with the help of simultaneous observations on a second apparatus, in which the upper end of the thread remains unaffected: it is hardly necessary to recall, that the distance between the two apparatus must be large enough, so that they cannot significantly interfere with each other.

In order to show how great a refinement these sorts of observations permit, we introduce an example from the daily log book. On September 22, 1832, pending the collimation error, the following declinations u and angle v[179] were observed:

Experiment	Time	First Needle		Second Needle
		u	v	u
I	9h 33' am	+ 0° 4' 19.5"	300°	+ 0° 2' 12.1"
II	9h 57'	– 0° 0' 19.6"	240°	+ 0° 1' 37.7"
III	10h 16'	– 0° 4' 40.5"	180°	+ 0° 1' 18.8"

Hence the declinations of the first needle, related to the position of the first observation, are as follows:

[177] [Note by AKTA:] In German: *Torsionskraft*.
[178] [Note by AKTA:] The constant Θ is called the torsion coefficient of the thread or its torsion constant.
[179] [Note by CFG:] Both elements expand from left to right.

I	$u = 0^o\ 4'\ 19.5''$	$v = 300^o$
II	$+0^o\ 0'\ 14.8''$	240°
III	$-0^o\ 3'\ 47.2''$	180°

From this emerges the value of the fraction $\frac{TM}{\Theta}$ from the combination of observations

I and II 881.7
II and III 891.5
I and III .... 886.6.

The daily changes in the magnetic declination are decreased by the torsion in the proportion of 1 to $\frac{n}{n+1}$ where $\frac{TM}{\Theta}$ is set $= n$, a change which, if we use threads of such low torsion as the foregoing example shows, can be considered as insignificant. So far as the duration of the (infinitely small) oscillations, it can easily be concluded from the dynamic principles, that it is decreased by the torsion in proportion of 1 to $\sqrt{\frac{n}{n+1}}$. Actually, this relates to the case where $v = 0$. The formulas would hold true, however, in general, if we posited $\frac{TM\cos u^o}{\Theta}$ wherein we denote the value of u by u^o, which corresponds to the equilibrium: but the difference will certainly be insignificant.

7.9

The coefficient Θ depends essentially on the length, the thickness, and the material of the thread, and additionally in metallic threads on the temperature, in satin threads on the humidity: in contrast, it seems to depend not at all, in the former case (perhaps also in the latter, if they are single threads) on the weight they bear. The situation is different when the satin threads are multiplex, as they must be to hold heavy needles: in this case Θ increases with the suspended weight, yet it remains far smaller than the value of Θ for a metallic thread of exactly the same length and load capacity. Thus, through a very similar method to that developed in the previous Section (although with a different thread and a different needle), the value of $n = 597.4$ is found, while the thread holds a needle with the usual equipment alone, where the total weight was 496.2 g; in contrast = 424.8, when the weight was increased to 710.8 g, or in the first case it was $\Theta = 0.0016740TM$, in the second case $\Theta = 0.0023542TM$. The thread, whose length was 800 *mm*, is composed of 32 individual threads,[180] which individually hold almost 30 *g* securely and are arranged so that they undergo an equal tension. Moreover, it is probable, that the value of Θ consists of a constant element and an element proportional to the weight, and that the constant part will equal the sum of the values of Θ for the individual simple threads. Under this hypothesis (which as of now is not adequately substantiated by experiments), 0.0001012 TM is found as the constant value for the example above, and thus 0.00000316 TM as the value of Θ for a simple thread. With recourse to the soon to develop value of TM, it will be calculated from this hypothesis, that the resistance of a single thread, wound around a curve equal to the radius ($57^o\ 18'$), is equivalent to the gravity of a milligram pressing on the arm of a lever with the length of approximately 1/17 *mm*.

[180][Note by CFG:] Strictly speaking, these threads are not really simple, but merely the kind that are sold as un-spun.

7.10

If the oscillating body is a simple needle of regular form and homogeneous mass, the moment of inertia K can be calculated according to known methods. If, e.g., the body is a right-angled parallelepiped, whose sides are a, b, c, whose density $= d$ and whose mass is q, thus $= abcd$, the moment of inertia will be $= (aa+bb)q/12$ in relation to an axis going through the midpoint and parallel to side c: and since in magnetic needles of such a shape, the side which is parallel to the magnetic axis, namely a, is usually far longer than b, it will, moreover, suffice for crude experiments to make $K = aaq/12$. But in more refined experiments, even if a simple needle is used, we would hardly be allowed the comfortable assumption of a completely homogeneous mass and a completely regular shape, and for those of our experiments in which, not a simple needle, but a needle involving more complex equipment, it is altogether impossible, to ascertain the state of affairs through such a calculation, and a different procedure was sought for the precise determination of the moment K.

To the needle was attached a wooden crossbeam, on which two equal weights hung, by means of which very fine spikes exerted pressure on points A and B of the beam: these points were on a horizontal straight line, in the same vertical plane as the axis of suspension, and were equidistant from it on both sides. If the mass of each of the two weights is denoted by p and the distance AB by $2r$, then by adding this apparatus the moment K will be increased by the magnitude $C + 2prr$, where C is the sum of the moment of the beam in relation to the axis of suspension, and the moment of the weights in relation to the vertical axis through the spikes and the center of gravity. Hence, if the oscillations of the un-weighted needle and the needle weighted at two different distances, are observed, namely, for $r = r'$ and $r = r''$, and the duration of oscillations (after they are reduced to infinitely small amplitudes and freed from the effect of torsion) respectively $= t$, t', t'', are found, then from combining the equations:

$$TMtt = \pi\pi K$$

$$TMt't' = \pi\pi\,(K + C + 2pr'r')$$

$$TMt''t'' = \pi\pi\,(K + C + 2pr''r'')$$

the three unknowns TM, K and C can be determined. We will attain still greater precision, if, for several values of r, namely $r = r'$, r'', r''' and so forth, we observe the associated duration of oscillations t', t'', t''', and so forth, and, according to the method of least squares, determine the two unknowns x and y such that:

$$t' = \sqrt{\frac{r'r' + y}{x}}$$

$$t'' = \sqrt{\frac{r''r'' + y}{x}}$$

$$t''' = \sqrt{\frac{r'''r''' + y}{x}}$$

and so forth.

By this means we obtain:

$$TM = 2\pi\pi px$$

$$K + C = 2py\ .$$

Regarding this method, the following is to be observed:

I. If a not too smooth needle is used, it suffices, to simply place the wooden beam on it. If, however, the surface is very smooth, so that friction cannot retard the sliding of the beam, it is necessary, so that the whole apparatus may move like a single fixed body, to connect the beam more securely to the rest of the apparatus. In both cases, care should be taken, that points A and B are located with sufficient precision in a horizontal straight line.

II. Since the ensemble of such experiments requires several hours, the variability of the terrestrial magnetism within this time span, at least if the greatest precision is desired, cannot be neglected. Hence, before the elimination is undertaken, the observed times must be reduced to a constant value of T, e.g. to the mean value, corresponding to the first experiment. For this purpose, simultaneous observations of another needle (just as in Section 8) are necessary: if these observations have as the duration of *one* oscillation for the mean time of the single experiment respectively $= u,\ u',\ u'',\ u'''$ and so forth, then instead of using the observed values $t',\ t'',\ t'''$ and so forth for the calculation, $\frac{ut'}{u'},\ \frac{ut''}{u''},\ \frac{ut'''}{u'''}$, and so forth respectively are used.

III. A similar comment holds true with regard to the variability of M, which comes from the change in temperature, if that occurs during the experiment. But it is clear that the just-described reduction already includes this improvement in and of itself, if each of the two needles is subject to the same temperature, and is influenced in the same way by such a change.

IV. If the task is only to ascertain the value of TM, clearly the first experiment is superfluous. Yet it will be useful to conjoin to the experiment conducted with a weighted needle another with an un-weighted needle, in order for the value of K to be obtained at the same time, so that it can be taken as a base for these experiments which are performed at another time with the same needle, since it is evident that this value remains constant, even when T and M undergo a change over time.

7.11

To better explain this method, we include here an example from the large quantity of applications. The experiment conducted on September 11, 1832, yielded the following Table:

Experiment	Simultaneous oscillations		
	of the first needle		of the second needle
	Load	One oscillation	One oscillation
I	$r = 180\ mm$	$24.63956''$	$17.32191''$
II	$r = 130\ mm$	$20.77576''$	$17.32051''$
III	$r = 80\ mm$	$17.66798''$	$17.31653''$
IV	$r = 30\ mm$	$15.80310''$	$17.30529''$
V	Without load	$15.22990''$	$17.31107''$

Times were observed on a clock which every day lost $14.24''$ mean time, and each of the two weights weighed 103.2572 g; the distances r in millimeters were determined with microscopic precision; the duration of an oscillation, which was ascertained from at least 100 oscillations (in the fifth experiment, even from 677 for the first needle), had already been reduced to infinitely small curves: moreover, these reductions were insignificant, because of the very small amplitude of the oscillations[181] which it is possible to apply to our apparatus without detriment to the greatest precision. We wanted to reduce these oscillation times, first to the mean value of TM, which took place during the fifth experiment, under application of the rules in Paragraph II above;[182] then to the values, which would have resulted without torsion, through multiplication by $\sqrt{\frac{n+1}{n}}$ where n in the four first experiments $= 424.8$, in the fifth experiment $= 597.4$ (compare Section 9); finally to the mean solar time through multiplication by $\frac{864.00}{86385.76}$; our results were:

$$\begin{array}{c} \text{I. } 24.65717'' = t' \text{ for } r' = 180\ mm \\ \text{II. } 20.79228'' = t'' \text{ for } r'' = 130\ mm \\ \text{III. } 17.68610'' = t''' \text{ for } r''' = 80\ mm \\ \text{IV. } 15.82958'' = t'''' \text{ for } r'''' = 30\ mm \\ \text{V. } 15.24515'' = t \text{ for the unloaded needle.} \end{array}$$

If we take the second, the millimeter and the milligram for the units of time, distance and mass, so that $p = 103257.2$, we infer from combining the first experiment with the fourth:

$$TM = 179,641,070\ ; \qquad K + C = 4,374,976,000\ ,$$

and then from the fifth experiment

$$K = 4,230,282,000 \qquad \text{and likewise} \qquad C = 144,694,000\ .$$

When, however, it is desired to take into account all the experiments, the method of least squares is the most convenient in the following way. We proceed from the approximate values of the unknowns x and y, which emerge from the combination of the first and fourth experiments, and, designating the still-to-be- added corrections by ξ and η, we say:

$$x = 88.13646 + \xi$$

181 [Note by CFG:] Thus the amplitude of the oscillations in the first experiment was $0^\circ\ 37'\ 26''$ at the beginning, $0^\circ\ 28'\ 34''$ at the end; in the fifth experiment $1^\circ\ 10'\ 21''$ at the beginning, after 177 oscillations $0^\circ\ 45'\ 35''$, and after 677 oscillations $0^\circ\ 6'\ 44''$.

182 [Note by AKTA:] See page 62.

$$y = 21184.85 + \eta \ .$$

From this are derived as calculated values the times t', t'', t''', t'''' by familiar methods:

$$t' = 24.65717 - 0.13988\xi + 0.00023008\eta$$

$$t'' = 20.78731 - 0.11793\xi + 0.00027291\eta$$

$$t''' = 17.69121 - 0.10036\xi + 0.00032067\eta$$

$$t'''' = 15.82958 - 0.08980\xi + 0.00035838\eta \ ,$$

whose comparison with the observed values, treated according to the method of least squares, yields the results:

$$\xi = -0.03230 \ ; \qquad \eta = -12.38 \ ;$$

$$x = 88.10416 \ ; \qquad y = 21172.47 \ .$$

From this finally results

$$TM = 179,575,250 \ , \qquad K + C = 4,372,419,000 \ ,$$

and then with addition of the first experiment:

$$K = 4,228,732,400 \ , \qquad C = 143,686,600 \ .$$

I next present a comparison of the times calculated from the corrected values of the magnitudes x, y, with the observed values.

Experiment	Calculated time	Observed time	Difference
I	24.65884″	24.65717″	+0.00167″
II	20.78774	20.79228	−0.00454
III	17.69046	17.68610	+0.00436
IV	15.82805	15.82958	−0.00153

In Göttingen, we set the length of the simple seconds pendulum = 994.126 mm, hence the weight, measured by that unit of the accelerating forces, which underlies the preceding calculations, = 9,811.63; thus if we prefer to take gravity itself as the unit, then $TM = 18,302.29$: this number expresses the number of milligrams, whose pressure under the influence of gravity on a lever of the length of one millimeter is equivalent to the force with which the terrestrial magnetism seeks to turn that needle around the vertical axis.

7.12

After we have completed the determination of the product of the horizontal components T of the terrestrial magnetic force into the magnetic moment M of the given needle,[183] we proceed to the second part of the experiment, namely, to the determination of the quotient M/T. We will accomplish this by comparing the action of this needle on another needle with the effect of terrestrial magnetism on the very same needle; specifically, this will be observed, as was already discussed in the introduction, either in a state of motion or in a state of equilibrium; we have repeatedly investigated each of the two methods: but for several reasons the latter is preferable by far to the former, and thus we will confine the investigation here to that one, while the first can be dealt with in a similar way without difficulty.

7.13

The conditions of equilibrium in a movable body, on which arbitrary forces act, can easily be brought together in one single formula by means of the principle of virtual displacements: namely, the sum of the products of any one force multiplied into the projection of an infinitely small displacement of its working point on the direction of force, must be so constituted that it can attain a positive value for no virtual movement, i.e., a movement complying with the conditions, so that, if the virtual movements are possible collectively in opposite directions, that aggregate, which we wish to designate $d\Omega$, is $= 0$ for any virtual movement.

The movable body we consider here is the magnetic needle, which is connected at a point G to a revolving thread, fastened at the top. This thread merely prevents the distance of point G from the fixed end of the thread from becoming greater than the length of the thread, so that here, too, as in the case of a completely free body, the position of the body in space depends on six variables and further the equilibrium itself on six conditions. Since, however, the solution of the problem is only to serve to determine the fraction M/T, it suffices to consider that virtual movement, which occurs in the rotation around the vertical axis intersecting G; and it is evident that such an axis can be considered as fixed, and only the angle between the vertical plane, in which the magnetic axis is located, and the magnetic meridian plane, as variable. We will take this angle from the north side of the meridian toward the east and designate it u.

7.14

We will conceive the volume of the movable magnetic needle as divided into infinitely small elements, letting the coordinates of an arbitrary element be x, y, z, while e is the element of free magnetism contained in it. We place the origin of the coordinates in the arbitrary point h inside the needle on the vertical line going through G; the axis of the coordinates x and y is to be horizontal, the former in the magnetic meridian directed toward the north, the latter toward the east; the coordinate z is directed upwards. Then the effect of the terrestrial magnetism on the element e yields the element $Tedx$ of $d\Omega$.

In a similar way, the volume of the second fixed needle is divided into infinitely small elements, and any element may conform to the coordinates X, Y, Z, and the amount E of free magnetism; finally, r would be $= \sqrt{(X-x)^2+(Y-y)^2+(Z-z)^2}$. Under this

[183][Note by AKTA:] In German: *das magnetische Moment M der gegebenen Nadel.*

condition, the action of element E on element e yields the contribution $eEdr/r^n$ to the sum $d\Omega$, if the power r^n of the distance r is assumed to be inversely proportional. If we denote by N the value of u which corresponds to the untwisted state of the thread, then the moment of the torsional force of the thread can be expressed as $\Theta(N-u)$. This force can be conceived in such a way, as if the tangential force $= \Theta(N-u)/D$ acted on each end of a horizontal diameter of the thread placed through G, where D denotes this diameter, and it is easily seen, that from this as element of the sum $d\Omega$ is derived: $\Theta(N-u)du$.

The weight of the particles of the needle clearly does not contribute to the sum $d\Omega$, since only u is variable; therefore we have the equation:

$$d\Omega = \sum Tedx + \sum \frac{eEdr}{r^n} + \Theta(N-u)du \ ,$$

in which the summation in the first term relates to all elements e, in the second to all combinations of the individual e with the individual E. Hence it is clear, that the condition of stable equilibrium consists in

$$\Omega = \sum Tex - \sum \frac{eE}{(n-1)r^{n-1}} - \frac{1}{2}\Theta(N-u)^2$$

becoming a maximum.

7.15

It suits our purpose, to always construct the experiments so that the magnetic axis of each of the two needles is horizontal, and that both needles are approximately at the same height; hence we wish to confine further calculations to these preconditions.

We will relate the coordinates of the points of the first needle to fixed axes in the needle, which also still intersect point h; that is, the first axis would lie in the direction of the magnetic axis, the second horizontal and to the right of the first, the third vertical and directed upward; the coordinates of the element e in relation to these axis would be a, b, c. Likewise A, B, C would be the coordinates of element E in relation to similarly fixed axes in the second needle, which intersect this needle at a point H: we select this point close to the center of the needle and at the same height as point h.

The position of point H would of course be most conveniently determined by the distance from point h and the direction of the straight line associated with it, if it is a matter of only *one* experiment: since, however, for our purpose several experiments are always necessary, which are related to different positions of point H, all of which certainly lie in the same straight line, yet not necessarily in a line through point h, it is better to construct the demonstration from the very beginning, in such a way that the system of such experiments depends on one single unknown. Hence we wish to relate point H to an arbitrary point h' in the same horizontal plane, which lies near h and whose coordinates may be α, β, 0, and wish to make the distance $h'H = R$ and the angle of the lines $h'H$ with the magnetic meridian $= \psi$. If we now denote the angle of the magnetic axis of the second needle with the magnetic meridian by U, then we will have:

$$x = a\cos u - b\sin u$$

$$y = a\sin u + b\cos u$$

$$z = c$$

$$X = \alpha + R\cos\psi + A\cos U - B\sin U$$

$$Y = \beta + R\sin\psi + A\sin U + B\cos U$$

$$Z = C \; .$$

Thus everything is in place for developing the sum $d\Omega$ and the fraction $\frac{d\Omega}{du}$, which has to disappear for the equilibrium state.

7.16

First, will

$$\sum Tex = T\cos u \sum ae - T\sin u \sum be = mT\cos u \; ,$$

if we denote by m the moment of the free magnetism of the first needle $\sum ae$, since $\sum be$ is $= 0$: the element of $\frac{d\Omega}{du}$, which derives from the first term of Ω, will be $= -mT\sin u$.

If, for the sake of brevity, we say:

$$k = \alpha\cos\psi + \beta\sin\psi + A\cos(\psi - U) + B\sin(\psi - U) - a\cos(\psi - u) - b\sin(\psi - u)$$

$$\begin{aligned} l = [\alpha\sin\psi - \beta\cos\psi + A\sin(\psi - U) - B\cos(\psi - U) - a\sin(\psi - u) \\ - \; b\cos(\psi - u)]^2 + (C - c)^2 \; , \end{aligned}$$

thus rr will be $= (R + k)^2 + l$.

Since for exploitable experiments R must be far larger than the dimensions of each of the two needles, the magnitude $\frac{1}{r^{n-1}}$ can be expanded into the quickly convergent series:

$$\begin{aligned} R^{-(n-1)} - (n-1)kR^{-n} + \left(\frac{nn - n}{2}kk - \frac{n-1}{2}l\right) R^{-(n+1)} \\ - \left(\frac{1}{6}(n^3 - n)k^3 - \frac{1}{2}(nn-1)kl\right) R^{-(n+2)} + \ldots \; , \end{aligned}$$

whose law, if it were worth the effort, could be easily specified. The individual terms of the sum $\sum \frac{eE}{r^{n-1}}$, which derives from inserting the values of the magnitudes k and l, will contain a factor of the form:

$$\sum eEa^{\lambda}b^{\mu}c^{\nu}A^{\lambda'}B^{\mu'}C^{\nu'} \; ;$$

this is equal to the product of the factors $ea^\lambda b^\mu c^\nu$ and $EA^{\lambda'}B^{\mu'}C^{\nu'}$, which respectively depend on the magnetic state of the first and second needles. Taking this into consideration, what we may establish confines itself to the equations:

$$\sum e = 0\,, \qquad \sum ea = m\,, \qquad \sum eb = 0\,, \qquad \sum ec = 0\,,$$

$$\sum E = 0\,, \qquad \sum EA = M\,, \qquad \sum EB = 0\,, \qquad \sum EC = 0\,,$$

where we denote by M the moment of the free magnetism of the second needle. In the special case, that the shape of the first needle (the moveable one) and the distribution of magnetism in the longitudinal direction are symmetrical, so that two elements always correspond, for which a and e have equivalent opposite values b and c, then, as soon as the midpoint coincides with point h, $\sum ea^\lambda b^\mu c^\nu$ will always $= 0$ for a direct value of the number $\lambda + \mu + \nu$, and similarly for the second needle, if the shape and the distribution of magnetism is symmetrical in relation to point H. Hence, in general, in the sum $\sum \frac{eE}{r^{(n-1)}}$ the coefficients of the powers $R^{-(n-1)}$ and R^{-n} disappear; in the special case, where each of the two needles is symmetrically shaped and symmetrically magnetized, while at the same time the midpoint of the first, h and h', coincide, and likewise the midpoint of the second and H coincide, then the coefficients of the powers $R^{-(n+2)}$, $R^{-(n+1)}$, $R^{-(n+6)}$ and so forth will also disappear; every time those conditions occur very approximately, they must at least be very small. The major term, which is derived from the elaboration of the second element of Ω, namely from $-\sum \frac{eE}{(n-1)r^{(n-1)}}$ will be:

$$= -\frac{1}{2}R^{-(n+1)}\left(n\sum eEkk - \sum eEl\right)$$

$$= mMR^{-(n+1)}\left(n\cos(\psi - U)\cos(\psi - u) - \sin(\psi - U)\sin(\psi - u)\right)\,.$$

From this it follows, that the element of $d\Omega/du$, which corresponds to the influence of the second needle, can be expressed by the following series:

$$fR^{-(n+1)} + f'R^{-(n+2)} + f''R^{-(n+3)} + \ldots\,,$$

in which the coefficients contain rational functions of the cosine and sine of the angles ψ, u, U and the magnitudes are α, β, and beyond that, contain constant magnitudes, which depend on the magnetic state of the needle; specifically, they will be:

$$f = mM\left(n\cos(\psi - U)\sin(\psi - u) + \sin(\psi - U)\cos(\psi - u)\right)\,.$$

The complete elaboration of the following coefficients f', f'' and so forth is not necessary for our purpose; it suffices to remark that

1) in the case of complete symmetry, the just-indicated coefficients f', f'' and so forth disappear;

2) if the remaining magnitudes remain unchanged and is increased by two right angles (or, the same thing, if the distance R is taken from the same, backwards-extended straight line on the other side of point h'), the coefficients f, f'', f'''' and so forth preserve their values, while f', f''', f''''' and so forth take on opposite values, or that the series turns into

$$fR^{-(n+1)} - f'R^{-(n+2)} + f''R^{-(n+3)} - \ldots\,;$$

this is easily inferred from the fact that by means of this change in ψ, k becomes $-k$, but l is not transformed.

7.17

The condition that the movable needle not be revolved around the vertical axis by the combination of forces is thus summed up in the following equation:

$$0 = -mT \sin u + fR^{-(n+1)} + f'R^{-(n+2)} + f''R^{-(n+3)} + \ldots - \Theta(u - N) .$$

Since it can easily be effected, that the value of N, if not $= 0$, is at least very small, and also u, for the experiment at hand, remains within narrow limits, then, without having to fear significant error, for the term $\Theta(u - N)$ one can use $\Theta \sin(u - N)$, all the more so, as $\frac{\Theta}{mT}$ is a far smaller fraction. Let u^{o} be the value of u, which corresponds to the equilibrium of the first needle in the absence of the second, or let

$$mT \sin u^{\text{o}} + \Theta \sin(u^{\text{o}} - N) = 0 ;$$

from this easily follows

$$mT \sin u + \Theta \sin(u - N) = (mT \cos u^{\text{o}} + \Theta \cos(u^{\text{o}} - N)) \sin(u - u^{\text{o}}) ,$$

where, in place of the first factor, $mT + \Theta$ can be adopted without reservation. Thus our equation becomes:

$$(mT + \Theta) \sin(u - u^{\text{o}}) = fR^{-(n+1)} + f'R^{-(n+2)} + f''R^{-(n+3)} + \ldots$$

If we keep only the term $fR^{-(n+1)}$, the solution is within reach, namely, we have

$$\tan(u - u^{\text{o}}) = \frac{mM \left(n \cos(\psi - U) \sin(\psi - u^{\text{o}}) + \sin(\psi - U) \cos(\psi - u^{\text{o}})\right) R^{-(n+1)}}{mT + \Theta + mM \left(n \cos(\psi - U) \cos(\psi - u^{\text{o}}) - \sin(\psi - U) \sin(\psi - u^{\text{o}})\right) R^{-(n+1)}}$$

where in the denominator of the element, which contains the factor $R^{-(n+1)}$, we can suppress or affirm, with just the same correctness:

$$\begin{aligned} \tan(u - u^{\text{o}}) = \frac{mM}{mT + \Theta} &[n \cos(\psi - U) \sin(\psi - u^{\text{o}}) \\ &+ \sin(\psi - U) \cos(\psi - u^{\text{o}})] R^{-(n+1)} = FR^{-(n+1)} . \end{aligned}$$

However, when we want to take into account the further terms, then it is clear, that $\tan(u - u^{\text{o}})$ can be expanded into a series of the following form:

$$\tan(u - u^{\text{o}}) = FR^{-(n+1)} + F'R^{-(n+2)} + F''R^{-(n+3)} + \ldots ,$$

in which, as a slight reflection shows, the coefficients F, F', F'' and so forth up to the coefficients of the power $R^{-(2n+1)}$ respectively result inclusively from

$$\frac{f}{mT+\Theta}\,, \qquad \frac{f'}{mT+\Theta}\,, \qquad \frac{f''}{mT+\Theta}\,, \qquad \text{and so forth}\,,$$

by means of changing u into u° from the following term on, however, new elements will enter, which for our purpose we need not pursue more precisely. For the rest, it is manifest that $u-u^{\circ}$ can be expanded into a series of a similar form, which, up to the power $R^{-(3n+2)}$, coincide with the series for $\tan(u-u^{\circ})$.

7.18

It is now clear that, if the second needle is set up at different points on the same straight line, so that ψ and U retain their value, while R alone is changed, and the deviation of the moveable needle from the equilibrium position, whereby the second needle is displaced, namely, the angle $u-u^{\circ}$ is observed, from this it follows that the values of the coefficients F, F', F'' and so forth, as many as are still significant, can be ascertained through elimination; by this means we will obtain the equation:

$$\frac{M}{T}=\left(1+\frac{\Theta}{Tm}\right)\frac{F}{n\cos(\psi-U)\sin(\psi-u^{\circ})+\sin(\psi-U)\cos(\psi-u^{\circ})}$$

in which the value of the magnitude $\frac{\Theta}{Tm}$ can be found by the method we demonstrated in Section 8. For a more convenient demonstration, however, it will be useful to observe the following:

I. In place of the comparison of u with u° it is preferable to compare the two opposite deviations with each other, by reversing the position of the second needle, namely, so that R and ψ remain unchanged and the angle U is increased by two right angles. If the values of u corresponding to these positions are denoted by u' and u'', then in the case of complete symmetry u'' precisely $=-u'$, if u° were $=0$. But it is superfluous, to anxiously keep to these conditions, since it is clear that u' and u'' are determined by similar series, in which the first terms have *precisely* opposite values, and further also $\frac{1}{2}(u'-u'')$, likewise $\tan\frac{1}{2}(u'-u'')$ by a similar series, in which the coefficient of the first term is *precisely* $=F$.

II. It will be still better, always to combine every four experiments, also after the angle is changed by two right angles or the distance R has been taken on the other side. If the two latter experiments correspond to the values u''' and u'''', then the difference $(u'''-u'''')/2$ will be expressed by a similar series, whose first term likewise will have a coefficient $=F$. It should be noted (which follows readily from the foregoing) that, if n were an odd number, the coefficients F, F'', F'''', and so forth would be precisely the same to infinity in any series for $u'-u^{\circ}$ and $u'''-u^{\circ}$, and the coefficients F', F''', F''''' and so forth would be precisely opposite to infinity, and the same for $u''-u^{\circ}$ and $u''''-u^{\circ}$, so that in the series for $u'-u''+u'''-u''''$ the alternate term would disappear. But in the case of actual reality, where $n=2$, such a relationship between the series for $u'-u^{\circ}$ and $u'''-u^{\circ}$ is, generally speaking, not strictly present, since, for the power R^{-6}, coefficients which are not precisely opposite already arise; however, it can be shown, that for this term as well a complete cancellation occurs in the combination $u'-u''+u'''-u''''$, so that $\tan\frac{1}{4}(u'-u''+u'''-u'''')$ has the form:

$$LR^{-3}+L'R^{-5}+L''R^{-7}+\ldots$$

or, more generally, if we leave the value of n undetermined for the time being, the following form:

$$LR^{-(n+1)} + L'R^{-(n+3)} + L''R^{-(n+5)} + \dots ,$$

where $L = F$.

III. It will be useful to choose the angles ψ and U in such a way that small errors occurring in the process of measurement itself, do not significantly change the value of F. For this purpose, the value of U for a given value of ψ must be posited in such a way, that F will be a maximum; namely, it must be

$$\cot(\psi - U) = n \tan(\psi - u^o) .$$

Then

$$F = \pm \frac{mM}{mT + \Theta} \sqrt{(nn \sin(\psi - u^o)^2 + \cos(\psi - u^o)^2)} .$$

The angle ψ is to be chosen, however, such that this value of F becomes either a maximum or a minimum: the former occurs for $\psi - u^o = 90^o$ or 270°, in which case

$$F = \pm \frac{nmM}{mT + \Theta} ,$$

the latter occurs for $\psi - u^o = 0$ or 180°, where

$$F = \pm \frac{mM}{mT + \Theta} .$$

7.19

Hence two methods are available which are best suited to carrying out our task. Their elements are shown in the following schema.

First Method.

The midpoint and the axis of the second needle lie in the straight line perpendicular to the magnetic meridian.[184]

Deflection	Position of the needle		Midpoint toward	North pole toward
$u = u'$	$\psi = 90^o$	$U = 90^o$	East	East
$u = u''$	$\psi = 90$	$U = 270$	East	West
$u = u'''$	$\psi = 270$	$U = 90$	West	East
$u = u''''$	$\psi = 270$	$U = 270$	West	West

Second Method.

[184][Note by CFG:] More precisely, to the vertical plane, which corresponds to the value $u = u^o$, i.e. in which the magnetic axis finds itself in equilibrium, when the second needle is not there. For the rest, in practice, the difference can be safely ignored, both because of the smallness and directly because of the relationships we discussed in Paragraph III above.

The midpoint of the second needle lies in the magnetic meridian.

Deflection	Position of the needle		Midpoint toward	North pole toward
$u = u'$	$\psi = 0°$	$U = 270°$	North	West
$u = u''$	$\psi = 0$	$U = 270$	North	East
$u = u'''$	$\psi = 180$	$U = 90$	South	West
$u = u''''$	$\psi = 180$	$U = 90$	South	East

If we set $\frac{1}{4}(u' - u'' + u''' - u'''') = v$ and

$$\tan v = LR^{-(n+1)} + L'R^{-(n+3)} + L''R^{-(n+5)} + \dots ,$$

then for the first method

$$L = \frac{nmM}{mT + \Theta} ,$$

for the second

$$L = \frac{mM}{mT + \Theta} .$$

7.20

From the theory of elimination it will easily be concluded, that the calculation is more imprecise because of the unavoidable errors in observation, the more coefficients must be determined by elimination. Therefore the method set forth in Section 18, II[185] is greatly to be prized, because it suppresses the coefficients $R^{-(n+2)}$, $R^{-(n+4)}$. In the case of complete symmetry, these coefficients would of course fall out by themselves, but it would be too uncertain to rely upon that case occurring. Moreover, a small deviation from the symmetry would have a far lesser influence in the first method than in the second, and if at least care is taken that point h', from which the distances will be measured, lies with sufficient precision in the magnetic meridian going through point h, there will scarcely be a significant difference between $u' - u''$ and $u''' - u''''$. However, things stand otherwise in the second method, especially if the apparatus requires an eccentric suspension. This method, whenever space does not permit observations from both sides, will always attain far less precision. Moreover, the first method is also especially preferable, because it yields a value of L twice as large as the second, since in the case of reality $n = 2$. If, by the way, one wants to discard, as much as possible in the second method, the term dependent on $R^{-(n+2)}$ in the case of the eccentric suspension, point h' should be chosen, so that the midpoint of the needle (for $u = u°$) lies in the center between h and h': the computation which showed this, however, I must omit for the sake of brevity.

7.21

In the foregoing computations, we have left the exponent n undetermined: during the days from June 24 to June 28, 1832, we carried out two series of experiments, under extension to

[185][Note by AKTA:] See page 70.

the greatest distances as the room permitted, through which it is shown most intelligibly, which values Nature demands. In the first series, the second needle (according to the method in Section 19) was placed in a straight line perpendicular to the magnetic meridian, in the second series, the midpoint of the needle was itself placed in the meridian. Here is an overview of these experiments, in which the distances R are expressed in fractions of meters and the values of the angle $(u' - u'' + u''' - u'''')/4$ are denoted for the first series by v, for the second by v'.

R	v	v'
1.1 m		1° 57′ 24.8″
1.2		1 29 40.5
1.3	2° 13′ 51.2″	1 10 19.3
1.4	1 47 28.6	0 55 58.9
1.5	1 27 19.1	0 45 14.3
1.6	1 12 7.6	0 37 12.2
1.7	1 0 9.9	0 30 57.9
1.8	0 50 52.5	0 25 59.5
1.9	0 43 21.8	0 22 9.2
2.0	0 37 16.2	0 19 1.6
2.1	0 32 4.6	0 16 24.7
2.5	0 18 51.9	0 9 36.1
3.0	0 11 0.7	0 5 33.7
3.5	0 6 56.9	0 3 28.9
4.0	0 4 35.9	0 2 22.2

Even a superficial glance shows that for larger values, the numbers in the second column nearly twice as big as the numbers of the third, and also the numbers in each row approximate the inverse of the cube of the distances, so that no doubt can remain as to the accuracy of the value $n = 2$. In order to confirm this law still further by means of specific experiments, we have dealt with all the numbers according to the method of least squares, from which the following values of the coefficients emerged:

$$\tan v = 0.086870R^{-3} - 0.002185R^{-5}$$

$$\tan v' = 0.043435R^{-3} + 0.002449R^{-5} \ .$$

The following overview shows the comparison of the values computed by this formula with the observed values.

Computed values.				
R	v	Difference	v'	Difference
1.1 m			$1^o\ 57'\ 22.0''$	$+2.8''$
1.2			1 29 46.5	−6.0
1.3	$2^o\ 13'\ 50.4''$	$+0.8''$	1 10 13.3	+ 6.0
1.4	1 47 24.1	+ 4.5	0 55 58.7	+ 0.2
1.5	1 27 28.7	−9.6	0 45 20.9	−6.6
1.6	1 12 10.9	−3.3	0 37 15.4	−3.2
1.7	1 0 14.9	−5.0	0 30 59.1	−1.2
1.8	0 50 48.3	+ 4.2	0 26 2.9	−3.4
1.9	0 43 14.0	+ 7.8	0 22 6.6	+ 2.6
2.0	0 37 5.6	+ 10.6	0 18 55.7	+ 5.9
2.1	0 32 3.7	+ 0.9	0 16 19.8	+ 4.9
2.5	0 19 2.1	−10.2	0 9 38.6	−2.5
3.0	0 11 1.8	−1.1	0 5 33.9	−0.2
3.5	0 6 57.1	−0.2	0 3 29.8	−1.0
4.0	0 4 39.6	−3.7	0 2 20.5	+ 1.7

7.22

The foregoing experiments were mainly undertaken with the intention of securing the law of magnetic action against any doubt, and further, of examining how many terms of the series are to be taken into account, and what degree of precision the experiments permit. They have shown that, if we do not make the distances smaller than four times the length of the needle, two terms suffice.[186] Furthermore, the differences which the computation has yielded, may in no way be simply held to be observational errors: several precautionary measures, from whose application a greater conformity may be expected, were at that time not yet in readiness. These include corrections for the hourly variability of the intensity of the terrestrial magnetism, of which heed must be taken in using another needle for comparison, according to the method we have spoken of in Section 10. However, so that one gets to know the value of the terrestrial magnetism, insofar as it can be inferred from these experiments, we append a synopsis of the remaining experiments in this group.

The value of the fraction $\frac{\Theta}{Tm}$ for the first needle and the thread on which it hangs, is ascertained by the method described in Section 8 to be $= \frac{1}{251.96}$.

From this comes the result:

$$\frac{M}{T} = 0.0436074 \ .$$

This number is based on the meter as the unit of length. If we prefer to take the millimeter, this number should be multiplied by the cube of 1,000, so that

$$\frac{M}{T} = 43,607,400 \ .$$

[186][Note by CFG:] The length of the needles used in these experiments was approximately 0.3 m; if we had tried to include the term R^{-7} in the computation, the precision would have been decreased rather than increased.

For the second needle, experiments were performed on June 28 resembling those we have described for another needle in Section 11; millimeters, milligrams, and the seconds of the mean sun time are taken as units, resulting in:

$$TM = 135,457,900 \ ,$$

and from this, by eliminating the magnitude M,

$$T = 1.7625 \ .$$

7.23

If experiments are performed to determine the absolute value T of the terrestrial magnetism, it is of great significance to take care that the entirety of these experiments be completed within not too long a time span, so that no significant change in the magnetic state of the needles used in the experiments is to be feared. It is recommended, that in observing the deviations of the movable needle, the first procedure in Section 20 alone be applied, after only two different distances are appropriately chosen, assuming that two terms of the series suffice. We choose as an example one of several applications of this method, namely, the one to which the most exacting care was devoted, by measuring the distances with microscopic precision.

The experiments were performed on September 18, 1832, with two apparatus, which we wish to denote by A and B, and specifically with three needles, denoted 1, 2, 3. Needles 1 and 2 are the same ones referred to as first and second in Section 11. The experiments are divided into two parts.

First the simultaneous oscillations of needle 1 in apparatus A and needle 2 in apparatus B were observed. As the duration of an oscillation, reduced to infinitely small amplitudes, resulted

for needle 1....... $15.22450''$
for needle 2....... $17.29995''$,

the former for 304 oscillations, the latter for 264 oscillations.

Next, needle 3 was suspended in apparatus A, while needle 1 was placed in the straight line perpendicular to the magnetic meridian, eastward and also west and on both sides in a double manner, and the deviation of needle 3 for the single position of needle 1 was observed. This experiment, which was repeated for two different distances R, gave the following values of the angle v, whose significance is the same as in Section 19 and Section 21:

$R = 1.2\ m$	$v = 3^o\ 42'\ 19.4''$
$R' = 1.6\ m$	$v' = 1^o\ 34'\ 19.3''$

During this experiment, too, the oscillations of needle 2 in apparatus B were observed. The mean time corresponded to the value, computed from 414 oscillations, of the oscillation duration for infinitely small curves = 17.29484.

The time periods were observed on a clock, whose daily retardation was $14.24''$. If M and m denote the moment of the free magnetism for needle 1 and 3, and Θ the torsion constant of the threat in apparatus A, while it held needle 1 or 3 (whose weight is almost identical), then we have:

$$\frac{\Theta}{TM} = \frac{1}{597.4} ,$$

as in Section 11,

$$\frac{\Theta}{Tm} = \frac{1}{721.6} ;$$

because needle 3 was more strongly magnetized than needle 1.

The moment of inertia of needle 1 was already known from earlier experiments (see Section 11), which had yielded: $K = 4228732400$, in which millimeter and milligrams were taken as the units.

The change in the thermometer in both rooms where the apparatuses were installed, was so slight during the entire period of the experiments, that it is superfluous to consider it.

We now wish to proceed to the computation of these experiments, in order to infer from it the intensity T of the terrestrial magnetism. The variation in the oscillations of needle 2 indicate a slight change in this intensity: hence, in order for us to speak of a definite value, we will reduce the observed duration of the oscillations of the first needle to the mean state of the terrestrial magnetism during the second part of the observations. This duration requires still another reduction on account of the retardation of the clock, and a third on account of the torsion of the thread. In this way the reduced duration of one oscillation produced:

$$= 15.22450 \times \frac{17.29484}{17.29995} \cdot \frac{86400}{86385.76} \cdot \sqrt{\frac{598.4}{597.4}} = 15.23500'' = t .$$

From this follows the value of the product

$$TM = \frac{\pi\pi K}{tt} = 179,770,600 .$$

The slight difference between this value and the one in Section 11 found on September 11, is to be ascribed to the change in the terrestrial magnetism and also to the change in the magnetic state of the needle.

From the observed deviations we derive:

$$F = \frac{R'^5 \tan v' - R^5 \tan v}{R'R' - RR} = 113,056,200 ,$$

if we take the millimeter as the unit, and thence

$$\frac{M}{T} = \frac{1}{2} F \left(1 + \frac{\Theta}{Tm}\right) = 56,606,437 .$$

Finally, the comparison of this number with the value of TM yields

$$T = 1.782088$$

as value of the intensity of the horizontal terrestrial magnetic force on September 18 at 5 o'clock.

7.24

The foregoing experiments were made at the Astronomical Observatory,[187] where the location for the apparatus was sought out in such a way that iron was kept away from the vicinity as much as possible. Nevertheless, it cannot be doubted, that the iron masses, which were abundantly distributed in the walls, windows, and doors of the building, indeed, even the iron components of the large astronomical instruments, in which magnetism is induced by the terrestrial magnetic force, exert an in no way insignificant effect on the suspended needles. The forces arising from this alter not only the direction, but also the intensity of the terrestrial magnetism by a small amount, and our experiments do not provide the pure value of the intensity of the terrestrial magnetism, but the value modified for the locus of apparatus A. As long as the iron masses remain in place and the elements of the terrestrial magnetism itself (namely, the intensity and direction) do not change very considerably, this modification must remain significantly constant, but what magnitude is reached, is indeed unknown up to now; nevertheless, I am scarcely inclined to believe, that this exceeds one or two hundredths of the total value. However, it should not be difficult to determine the magnitudes, at least approximately, by means of experiments, namely, through observation of the simultaneous oscillations of two needles, of which the one would be at the usual locus of observation, the other meanwhile at a rather great distance from the building and from other disturbing masses of iron, and which then would have to exchange places. But up to now it was not possible to carry out this experiment. The safest remedy, however, must be to construct a special building devoted to magnetic observations, which as a result of royal graciousness will soon be erected, and from whose construction iron is to be altogether excluded.

7.25

In addition to the experiments described, we have carried out many other similar experiments, even if we took less care with the earlier ones. It will, nevertheless, be of interest here to assemble the results in a Table, in which, however, those results are ignored, which before the installation of the more refined apparatus, were obtained by means of other, cruder adjuncts to needles of the most disparate dimensions, although all of them have provided at least an approximation of reality. Through repeated experiments, the following successive values of T resulted:

Number	Time, 1832	T
I	May 21	1.7820
II	May 24	1.7694
III	June 4	1.7713
IV	June 24-28	1.7625
V	July 23, 24	1.7826
VI	July 25, 26	1.7845
VII	September 9	1.7764
VIII	September 18	1.7821
XI	September 27	1.7965
X	October 15	1.7860

[187][Note by AKTA:] In German: *Sternwarte*.

Experiments V-IX were performed together in the same place, but I-IV in different places; experiment X is actually a composite, since the deviations were observed at the usual place, but the oscillations at another place. In experiments VII and VIII, an almost equal precision was applied; in experiments IV, V, VI and X, however, a somewhat lesser precision, and in experiments I-III a far lesser. In experiments I-VIII, in fact, different needles were used, although they had the same weight and the same length (the weight was between 400 g and 440 g); experiment X, in contrast, used a needle, whose weight was $1,062$ g and whose length was 485 mm. Experiment IX was only undertaken in order to see what degree of precision can be attained by means of a very small needle. The weight of the needle employed was only 58 g, otherwise, however, the precision was no less than in experiments VII and VIII. There exists no doubt, that the refinement of the observations will be significantly increased, when still heavier needles are used, e.g. needles weighing up to $2,000$ or $3,000$ g.

7.26

If the intensity T of the terrestrial magnetic force is expressed by a number k, then this is based on a certain unit V, namely, a force that is the same as the other force, whose connection with the other directly given units is certainly contained in the above, but in a somewhat complicated way: therefore, it will be worth the trouble, to demonstrate this connection here anew, in order to show with elementary clarity, what change the number k undergoes, if we start with other units instead of the original units.

To establish unit V, one must proceed from the unit of free magnetism M[188] and the unit of distance R, and we make V equal to the force exerted by M at distance R.

As unit M, we have assumed that quantity of magnetic fluid, which elicits, in an equally large quantity M at distance R, a motive force (or, if one prefers, a pressure), which is equivalent to that W, which serves as the unit, i.e., equivalent to the force, which the accelerating force A taken as the unit exerts on the mass P taken as the unit.

To establish unit A, two paths are available: namely, A can either be derived from a similar immediately given force, e.g. from the gravity at the locus of observation, or from the effect of A, which manifests itself in the movement of bodies. The second method, which we followed in our calculations, requires two new units, namely, the unit of time S and the unit of velocity C, so that that accelerating force is taken as the unit which, if it acts in time S, elicits the velocity C; finally, for the latter, that velocity is assumed, which corresponds to uniform movement through space R in time S.

Thus it is clear, that the unit V depends on three units, either R, P, A or R, P, S.

If we now assume that in place of units V, R, M, W, A, P, C, S, others are taken: V', R', M', W', A', P', C', S', which are connected with each other like the previous ones, and that the terrestrial magnetism will be expressed by the use of the measure V' through the number k', then it is to be investigated, how this relates to k.[189]

If we say:

$$V = vV'$$

$$R = rR'$$

[188] [Note by CFG:] It will scarcely be necessary to recall, that the earlier denotation given to the letters stops here.

[189] [Note by AKTA:] That is, it is to be investigated how this number k' relates to number k.

$$M = mM'$$

$$W = wW'$$

$$A = aA'$$

$$P = pP'$$

$$C = cC'$$

$$S = sS' ,$$

then v, r, m, w, a, p, c, s will be absolute numbers, and

$$kV = k'V' \qquad \text{or} \qquad kv = k' ,$$

$$v = \frac{m}{rr} ,$$

$$\frac{mm}{rr} = w = pa ,$$

$$a = \frac{c}{s} ,$$

$$c = \frac{r}{s} .$$

From the combination of these equations we find:

$$\text{I.} \qquad k' = k\sqrt{\frac{p}{rss}}$$

$$\text{II.} \qquad k' = k\sqrt{\frac{pa}{rr}} .$$

As long as we keep to the path we have followed in our observations, we are obliged to use the first formula: if, for example, we assume the meter and the gram as units instead of millimeters and milligrams, then r will be $= \frac{1}{1000}$, $p = \frac{1}{1000}$, thus $k = k'$; if the Paris line and the Berlin pound [are the units], then we will have: $r = \frac{1}{2.255829}$, $p = \frac{1}{467711.4}$, consequently

$$k' = 0.002196161k ,$$

and thus, e.g., experiment VIII gives the value $T = 0.0039131$.

If one prefers to follow another path, and assumes gravity as the unit of the accelerating force, then, for the Göttingen Observatory, a will $= \frac{1}{9811.63}$; then, if we keep the millimeter and milligram, the numbers k are to be multiplied by 0.01009554 and the alteration of the former unit is to be handled according to formula II.

7.27

The intensity of the horizontal terrestrial magnetic force T is to be multiplied by the secant of the inclination, in order to yield the total intensity. The fact that the inclination is variable in Göttingen and has undergone a diminution in the recent period, has been shown by the observations of *Humboldt*, who in the month of December 1805 found the value of $69^\circ\ 29'$, but in the month of September 1826 found $68^\circ\ 22'\ 26''$. Likewise I found on June 23, 1832, with the help of the same Inclinatorium which M. *Mayer* once used, $68^\circ\ 22'\ 52''$, which seems to signify a retardation in the decline; yet I am inclined to put less trust in this observation, not only because of the imperfection of the instrument, but also because of the circumstance, that the observation performed in the observatory was not adequately secured from disturbance by iron masses. However, these factors will also gain greater precision in future.

7.28

In this treatise, we have followed the generally accepted manner of explaining magnetic phenomena, not only because it is completely adequate, but also because it progresses in far simpler computations than the other view, which ascribes magnetism to galvano-electric currents around the particles of the magnetic body;[190] it was our intention, neither to confirm nor to refute this view, which, to be sure, recommends itself in several respects; this would have been inopportune, since the law of mutual effect among the elements of such currents does not yet seem to be sufficiently investigated. Whichever view is accepted, in the future as well, for magnetic and electromagnetic phenomena: the first theory [of currents] must always lead to the same results as the usual theory, and what is developed on the basis of the usual theory in the foregoing treatise, will be in a position to be changed merely in form, but not in essentials.

---

[190][Note by AKTA:] Gauss is referring to the hypothesis of molecular currents developed by André-Marie Ampère (1775-1836) between 1820 and 1827, see Chapter 5 (Ampère's Conception of Magnetism) of [AC15].

Ampère's masterpiece was published in 1826, [Amp26] and [Amp23]. There is a complete Portuguese translation of this work, [Cha09] and [AC11]. Partial English translations can be found at [Amp65] and [Amp69]. Complete and commented English translations can be found in [Amp12] and [AC15].

A huge material on Ampère and his force law between current elements can be found in the homepage *Ampère et l'Histoire de l'Électricité*, [Blo05].

Chapter 8

[Gauss, 1837] Introduction to the Results of the Observations Made by the Magnetic Association in the Year 1836

Carl Friedrich Gauss[191,192]

Being the First Annual Report of the Magnetic Association

Among the numerous phaenomena of terrestrial magnetism, with which we can only become acquainted by continued observations, accurately performed at various points of the earth's surface, none are in need of a more rigorously systematized cooperation of observers, than the *irregular variations* to which we find this force to be subject. It is sufficiently well known that the Variation, the Dip, and without doubt the Intensity also, (although with respect to the latter, which has but recently been admitted into the circle of inquiries, sufficient observations are still wanting) continually undergo changes — secular changes, which attract our attention only after long intervals of time, but which eventually become very considerable, — and periodical changes, varying according to the yearly and daily period. But for these regular changes, a rigorously systematized cooperation of observers, at various stations, is not essentially necessary, although highly desirable for the purpose of hastening the extension of our knowledge; in these points, every observer, even independently of others, may contribute useful additions.

Such, however, is not the case with respect to the irregular variations to which only of late years a larger share of attention has been devoted. Hiorter and Celsius observed, nearly a century ago, that during the appearance of an *aurora borealis*, the magnetic needle undergoes irregular and, frequently, very great oscillations; and this was subsequently confirmed by numerous observations made by others. Hence it might have been concluded, that the same forces which produce the phaenomenon of an aurora borealis act also at the same time upon the magnetic needle; and further, that this action must extend to very considerable distances,

[191] [Gau37b] with English translation in [Gau41b] and [GW66]; and French translation in [Gau38a].
[192] The Notes by A. K. T. Assis are represented by [Note by AKTA:].

since the northern lights are generally visible over a wide circuit. We obtain a still greater notion of the wide extension of the activity of these forces from the remark of M. Arago, that frequently on the same days, when he had observed at Paris violent disturbances of the regular movement of the magnetic needle, northern lights, not visible above the horizon of Paris, had been seen at distant places.

The irregularities in the phaenomena of terrestrial magnetism, the frequent occurrence of which had also been observed, especially by Humboldt in his numerous observations of the diurnal and horary oscillations of the magnetic needle, thus obtained a peculiar interest. Though the facts which had been remarked, neither proved that all irregular oscillations of the needle are contemporaneous with the northern lights, nor precluded the possibility, that many, perhaps most of them, have merely local causes, yet it was scarcely possible to mistake the evidence of the not unfrequent action, over a wide extent, of great natural forces, which, if they could not yet be investigated in their sources, offered at least a worthy object of natural inquiry, in respect to the relations of their activity and extent.

Superficial and merely accidental recognitions of such relations can bring us no nearer to this goal: in order to attain it, many such phaenomena must be contemporaneously followed up in accurate detail at numerous stations, and their time and magnitude closely ascertained and measured. For this purpose, however, previously concerted plans are essentially necessary among those observers who have suitable means at their disposal.

The celebrated philosopher to whom we are indebted for so many additions to our knowledge of terrestrial magnetism was also here the first to lead the way. M. von Humboldt caused to be erected in Berlin, towards the end of the year 1828, a small house, free from iron, — placed in it a variation compass constructed by Gambey, — and concerted with possessors of similar instruments at various places, some of which were very distant, regular observations of the magnetic variation on fixed days. Eight terms in the year were agreed upon, each of forty-four hours, during which the variation was to be noted from hour to hour; at some places observations were made within still narrower limits of time, viz. at every half-hour, or every twenty minutes. The details will be found in the nineteenth volume of Poggendorff's *Annalen der Physik*, p. 361;[193] and in the same journal are also the observations which, according to this agreement, were made at the appointed terms, in the years 1829 and 1830, in Berlin, Freiberg, St. Petersburg, Kasan, and Nicolajef, together with the graphical representations of three of them.

In the Göttingen magnetic observatory, which was built in the year 1833, and in which the magnetic apparatus is entirely different in construction from any previously employed, these term-observations were made for the first time on the 20th and 21st of March, 1834; corresponding observations were made in Berlin; but at Göttingen the observations were made every ten minutes, and in Berlin only every hour. Those at Berlin exhibited several considerable movements, which were found also in the Göttingen observations; while these latter exhibited in the intervening times a great number of movements which, of course, were entirely wanting in those made at Berlin. The question, whether the greater part of the fluctuations observed in Göttingen had been merely local, remained therefore still undecided.

The following term of the 4th and 5th of May, 1834, brought with it the decision. The intervening periods were more limited, the observations being made every five minutes, which gave to the results a considerably more definite character. No corresponding observations with Gambey's apparatus during this term, or in any subsequent ones, have been published. On the other hand, M. Sartorius, who had taken an active part in the March

[193][Note by AKTA:] [DH30].

term-observations at Göttingen, and who, being on the point of undertaking a journey of several years to Italy, had provided himself with an apparatus similar to the one at Göttingen, but of smaller dimensions, made with it careful and complete observations, at short intervals, during the May term, at Waltershausen, in Bavaria, about twenty German miles from Göttingen. A concordance surprisingly great was manifested, not only in the larger, but even in almost all the smaller oscillations, so that in fact nothing remained which could be justly ascribed to local causes.

During the three following terms, i.e. in June, August, and September, 1834, the observations were continued at Göttingen in exactly the same way; and the number of observers at other places, with apparatus either the same or of similar construction, was continually on the increase. Professor Encke, having become acquainted, from personal inspection, with the arrangements in Göttingen, ordered provisionally a similar apparatus of smaller dimensions for Berlin. M. Sartorius observed with his instrument during all the terms when circumstances allowed, viz. in June at Frankfort, and in September at Bramberg, in the province of Salzburg. Observations were also made in Leipzig, Copenhagen, and Brunswick, with instruments exactly resembling those of Göttingen. The result of the corresponding observations was quite similar to that above mentioned of the May term. Almost all the numerous movements observed at Göttingen occurred in the observations at other places, and although in varied relative magnitudes, yet with a concordance which did not admit of mistake.

In order to obtain further undeniable proof respecting this remarkable result, Professor Weber, being then at Leipzig, arranged that corresponding observations should be made at that place and at Göttingen, and certain hours of the forenoon, noon, and evening of the 1st and 2nd of October were fixed upon for the purpose. These observations, made by highly experienced observers, and with the greatest care, were published entire in Poggendorff's *Annalen der Physik*, vol. xxxiii. p. 426,[194] and elucidated by graphic representations. The necessity now became evident of observing the phaenomena at much shorter intervals than Humboldt had chosen. We observed during some of the appointed terms at intervals of three minutes, and some other observers did the same. As, however, several of the cooperators adhered to the five minute intervals, and as these in ordinary cases fully suffice, we subsequently, for the sake of uniformity, adopted this as a general rule. But as such small intervals render the labour incomparably more troublesome than the noting from hour to hour, especially in cases in which only a small number of persons can take part, it was necessary, in order to ensure the stability of the Association, to diminish both the number, and the duration of the terms. The number has since been fixed at six in a year, and the duration of each term at twenty-four hours. To each principal term two subordinate terms were added. Other details will be found in the sequel.

The observations have continued uninterruptedly, according to this plan, at Göttingen, and also at a constantly increasing number of other stations. Apparatus of the same or of similar construction to those in Göttingen, are employed in Altona, Augsburg, Berlin, Bonn, Brunswick, Breda, Breslau, Cassel, Copenhagen, Dublin, Freiberg, Göttingen, Greenwich, Halle, Kassan, Cracow, Leipzig, Milan, Marburg, Munich, Naples, St. Petersburg, and Upsala. From eight of these places no observations have yet come to our knowledge; and, in some others, the participation in the observations, from extrinsic circumstances, has not hitherto been uninterrupted and regular.

Some terms of the earlier period of the Association have been published in graphic rep-

[194][Note by AKTA:] [Gau34b].

resentations in Schumacher's *Astronomische Nachrichten*, and in Poggendorff's *Annalen der Physik*. The participation having so much increased, the time appeared to have arrived for taking into consideration a regular publication, in order that the abundant collection of fruitful facts might be made the common property of that portion of the public which is interested in these researches. What we now offer may be considered as the first annual report since the Association has attained a certain extent. From the year 1837, the results of each term will be made public as soon as they can be brought together in a sufficiently perfect manner.

The observations, and their graphical representation, will not merely be accompanied by those explanations and remarks which relate immediately to themselves; but we shall likewise add other memoirs, in which various subjects belonging to the wide field of terrestrial magnetism — the instruments, their use and manipulation, and various applications — will find a place.

With regard to the immediate object of the labours of our Association, the variations of the magnetic Declination, I may be allowed to add one more remark. If, as cannot be doubted, the two other elements of the terrestrial magnetic force, the Inclination and the Intensity, are subject to similar changes, the question may be asked, why such careful labour has been devoted to the first element, in preference, and hitherto exclusively?

The knowledge of the variations and the disturbances of the magnetic Declination possesses in fact a very great practical interest. To the mariner, and the surveyor, it must be of considerable importance to know the frequency and magnitude of the disturbances to which the compass is liable, even were it only to learn what degree of confidence he might place in its indications. For geodesical purposes the future progress of these inquiries may probably do much more. If it is once established that the irregular disturbances are never, or very seldom, merely local, — but that they constantly, or almost always, occur contemporaneously, and with almost equal magnitude, over great districts, the means are furnished to divest them almost entirely of any injurious practical effect. The surveyor need only make all his operations with the compass accurately according to time, and cause contemporaneous observations to be made at some other not very distant place; and it will be easy to eliminate the effects of these disturbances by comparison, just as travelling observers render their barometrical determinations of height independent of the irregular variations of the barometer, by comparative observations at fixed stations. Of course this has no reference to disturbances of the compass by mineralogical causes.

The preference given to the Declination over the other elements of terrestrial magnetism is less however to be ascribed to these motives than to the present state of our means. The investigation of the laws of nature has for the philosopher its own value and its own reward; and a peculiar charm surrounds the recognition of measure and harmony in that which at first sight appears wholly irregular. In following the constantly varying changes of the Declination, the apparatus at present employed leaves, as to certainty and precision, nothing more to wish; but the same cannot be said of the present means of observation of the other two elements. The time is therefore not yet come for including the latter in the circle of combined inquiry; as soon, however, as the means of observation shall be so far perfected, that we can recognise with certainty, follow with ease, and measure with accuracy, the variations, and chiefly the rapidly varying changes, in the other two elements of terrestrial magnetism, these variations will have the same claims on the united activity of natural inquirers as the variations of the declination now possess. We venture to hope that this day is not far distant.

Chapter 9

[Weber, 1837a] Remarks on the Arrangement of Magnetical Observatories, and Description of the Instruments to be Placed in Them

Wilhelm Weber[195,196]

The instruments with which the observations were made, which are to be mentioned in these pages, differ in many respects from all previously employed, and a more accurate knowledge of their construction is indispensable, in order to judge of the results obtained with them. It is true, that what the public have already been made acquainted with on this subject, in former memoirs and notices[197,198,199] might be sufficient; yet the perfect and accurate delineation of these instruments, which we shall give in this place, will render them easily understood, and will, besides, have the advantage that any clever artist can work from it with certainty. Instruments on the plan here represented have been made for Bonn, Dublin, Freiberg, Greenwich, Kasan, Milan, Munich, Naples, and Upsala, by Meyerstein of Göttingen;[200] and those for Göttingen, by Apel, and those for Cracow, Leipzig, and Marburg, by Breithaupt of Cassel, are almost perfectly similar. The description of the smaller instruments which have been employed at some places will be here omitted, since their use has been proved to be less proper, and only to be justified when local circumstances hinder the erection of larger apparatus. Nor will any mention be made of larger instruments, because if they are to fulfil all purposes, they require a proportionally larger place of reception

[195] [Web37a] with English translations in [Web41f], [GW66] and [GW39a]; and French translation in [Web38c].

[196] The Notes by E. Riecke, the editor of the second volume of Weber's *Werke*, are represented by [Note by ER:]; the Notes by Wilhelm Weber are represented by [Note by WW:]; the Notes by W. Francis, the translator of Weber's paper, are represented by [Note by WF:]; while the Notes by A. K. T. Assis are represented by [Note by AKTA:].

[197] [Note by WW:] In the memoir, *Intensitas vis magnetica terrestris ad mensuram absolutam revocata; auctore, C. F. Gauss, Gottingae*, 1833; further, in the *Göttingischen gelehrten Anzeigen*, 1832, p. 2011, 1835, p. 345, and in Schumacher's *Jahrbuche*, 1836, p. 1.

[198] [Note by ER:] Gauss' *Werke*, Vol. V, pp. 79, 293, 528 and 315.

[199] [Note by AKTA:] See Chapters 5, 6 and 7. See also [Gau35b], [Gau35a], [Gau36b] and [RR13, p. 238].

[200] [Note by AKTA:] Moritz Meyerstein (1808-1882). See [Hen04], [Hen05], [Hen07] and [Hen20].

than has hitherto been anywhere assigned to this object.

A long quadrangular *room*[201,202] which extends about eleven metres in the direction of the magnetic meridian, is best suited for the reception of magnetic instruments. It is not necessary that the side walls should be parallel with this meridian; they may form an angle with it, as is the case, for instance, at Göttingen, where they are in the direction of the astronomical meridian, which at present forms with the magnetic an angle of $18\frac{1}{2}$ degrees. The room must be well lighted, principally from the east and west, and more particularly at the end where the theodolite or the telescope, together with the scale, are to be placed for observation. The room should be protected from currents of air, for which purpose, a double door, and sometimes even double windows, are necessary; and there must be a solid foundation, upon which a *theodolite* and *clock* may be erected. It is also necessary that, from the place of the theodolite telescope, a distant object, the azimuth of which is known or may be accurately determined, may be seen through one of the windows. The floor in the neighbourhood of the instruments, i.e. near the centre of the room, must contain no iron, nor must any object containing that metal be brought near them. It is even desirable that the entire building, even as to its side walls and roof, should contain no iron; but it is unnecessary to be so cautious as to fear placing a clock, or a theodolite with steel pivots, at a distance of from five to six metres from the instrument. The influence of the steel parts, if they are magnetic, may be approximately deduced by calculation, and is found to be much too small to be sensible at those distances. Small pieces of iron outside the room have still less influence. If, however, there were in the neighbourhood large masses of iron, especially very long iron bars, (such as iron railings), although their influence would be very small, yet it should not be totally neglected. If they are at a distance of a hundred feet from the observatory, they offer no important impediment, at least if they are fixed. Such a locality is sufficient for measuring the Declination and Intensity, and also for observing their changes. Measurements of the Inclination may be performed in the same locality, but not, however, without interrupting the other observations. It therefore appears convenient, when circumstances permit, to assign a separate locality for measurements of the Inclination, which may be at no great distance from the first-named room. Where no absolute measurements are made, but only the changes of Declination observed at the fixed terms, such a room suffices, even should it contain much iron within and without its walls, provided that all the iron remains unmoved during the observations. The room of the Göttingen magnetic observatory is figured in Plate I;[203] and the ground-plan in Figure 1 of Plate II, and the site plan in Figure 2 of Plate II.

[201] [Note by WW:] The numbers appearing here refer to the following remarks about the individual parts of the magnetic observatory and the magnetic instruments.

[202] [Note by AKTA:] This footnote was not translated in [Web41f].

[203] [Note by AKTA:] The picture of this footnote appeared in the cover of the Magazine of Science, and School of Arts from October 22, 1842, Vol. 4, Number 30:

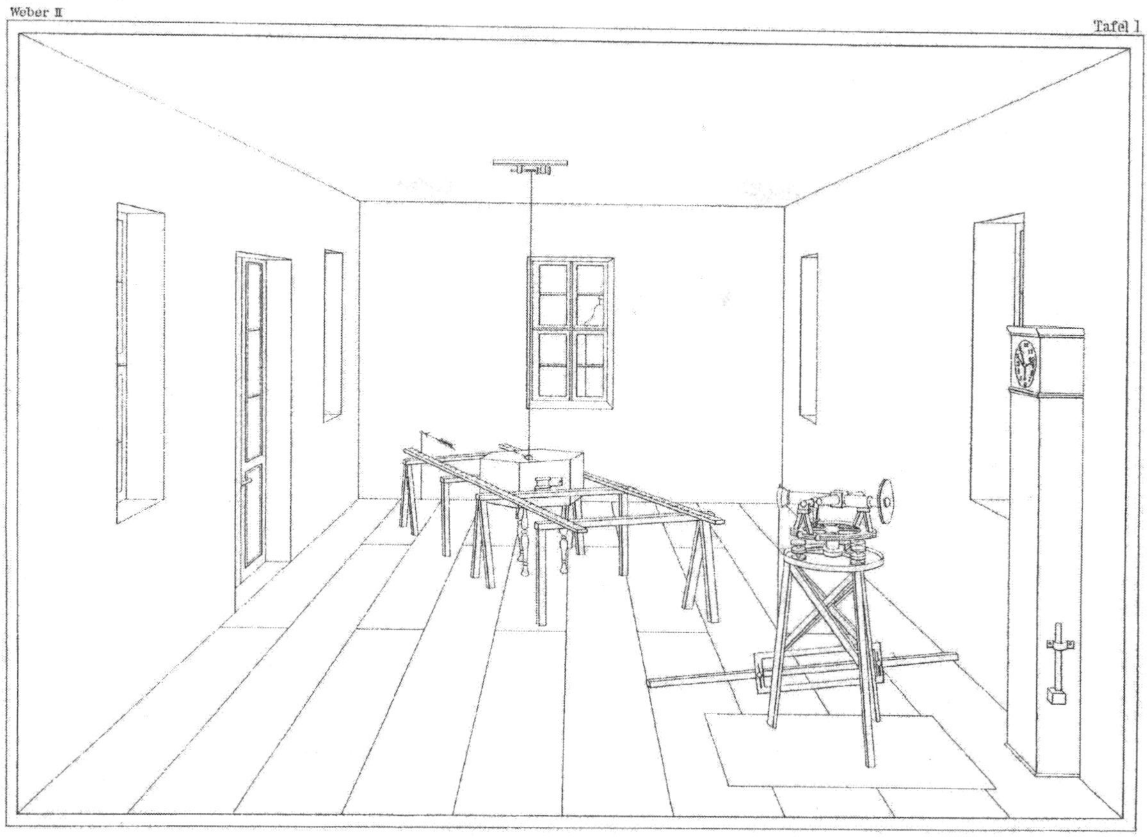

This Magazine also included a description of the Magnetic Observatory at Göttingen, [Ano42].

Weber II Tafel II

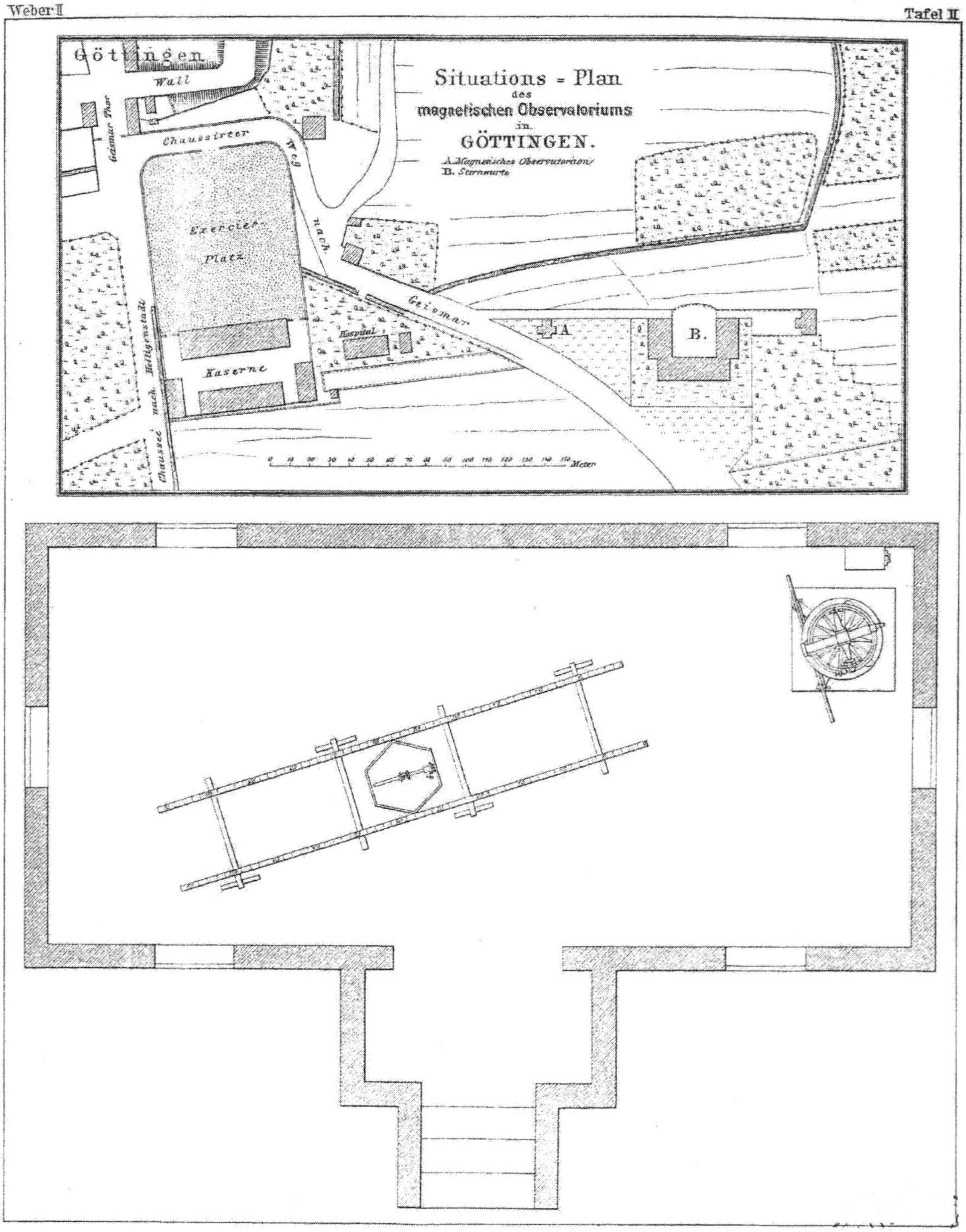

For the purpose of setting up the instruments, a line should be drawn on the floor representing the magnetic meridian, which line must pass nearly through the middle of the room, and terminate at the southern or northern end of it, where a firm foundation must be made for the theodolite and clock. When this foundation has been prepared, and the theodolite placed upon it, let a *scale* be first attached to the stand of the telescope, so that a plumb-line let fall from the object glass of the telescope passes freely before the scale. The scale must be horizontal and at right angles with the magnetic meridian; it must be capable of being raised or lowered at pleasure, and must be bisected by the magnetic meridian passing through the optical axis of the telescope. Next let fall a plumb-line from the ceiling to the floor, in such manner that the plane of the magnetic meridian passing through this plumb-line may contain the optical axis of the telescope; and when the *magnetometer*[204] is suspended by the plumb-line, the distances of the reflecting plane of the magnetometer (see *mirror* and *mirror-holder*) from the scale and from the telescope, may be together equal to the distance of the telescope from a point on the opposite wall, which is to serve as a mark, to which the telescope may be directed. At the point of the ceiling whence the plumb-line is let fall, the *suspender* of the magnetometer, together with the *elevating screw* and *suspension thread*, must be fixed. Let a weight be provisionally attached to the thread suspended from the elevating screw, in the manner of a plumb-line; adjust the suspender on the ceiling until the thread coincides with the plumb-line, making the length of the suspender parallel to the north or south wall of the room. After this, measure the height of the suspender, of the telescope, and of the scale from the ground. From the first height, subtract half the sum of the two latter, and form a thread of parallel fibres of raw silk (*Koconfäden*), whose length is equal to this difference, and which is sufficiently strong to carry the magnetometer, and one kilogramme of additional weight. The upper extremity of this thread is to be fastened to the screw, and the lower to the *stirrup*, (*Shiffchen*), in which the magnet bar is placed. A wide *box* is placed under the magnet bar, at the bottom of which are two cushions, upon which the magnet bar would fall, in case the raw silk fibres should break, without endangering the mirror attached to the front extremity of the magnet bar.

After these preparations the more accurate measurements may be commenced. These are:

1. To place the magnetic axis of the magnet in a horizontal direction, and the mirror perpendicular to it; or to measure the small angle which the axis of the mirror forms with the magnetic axis.

2. When the magnet is in its mean direction, to bring the force of torsion of the thread to zero, or to measure the small remaining torsion. (Vide seq. *torsion bar.*)

3. To determine the ratio of the moment of torsion of the thread, and the magnetic moment of the bar, in a deflection. (Vide seq. *stirrup* and *torsion circle.*)

4. To ascertain by measurement the place for the mark on the wall opposite to the telescope.

The apparatus is then ready for measurements of the Declination. These consist:

1. In the measurement of the azimuth of the mark.

2. In determining the values of the parts of the scale.

3. In observing the vibrations and elongations. (Vide seq. *quieting bar.*)

More accurate directions for the execution of all the measurements here mentioned will be given in the sequel.

For the measurements of Intensity *measuring scales* are required, by which the position of the *deflecting bar* is determined. These measuring scales may be laid horizontally and parallel to the magnetic meridian, on both sides of the box in which the magnetometer is included, in such manner that lines connecting the corresponding points of the two measuring scales shall be horizontal, and at right angles with the magnetic meridian. The scales should be placed at such height that the deflecting bar placed on them stand at an equal height with the vibrating bar. When this is not the case, the vertical distance between the deflecting bar situated on the measuring scales and the vibrating bar must be measured. The measuring scales must be about 5 to 6 metres in length, and should project an equal distance north and south beyond the magnetometer. If the width of the room allow, it is advantageous to add

[204][Note by AKTA:] The name *magnetometer* has been given by Gauss, [Gau36b, p. 332 of Gauss' *Werke*].

a third measuring scale horizontally and at right angles with the two former. It may pass under the box of the magnetometer, in such manner that it would be met by a plumb-line let fall from the middle between the centres of suspension and gravity of the vibrating bar. The measuring scales must be so arranged as to allow of displacing them longitudinally, in order to dispose them in such manner, that the deflecting bar, situated at corresponding points, in front and in rear, may produce the same amount of deviation. After these preparations, the measurement of the Intensity consists,

1. In determining the moment of inertia of the deflecting bar. (Vide seq. *weights* and *weight-holder.*)

2. In measuring the time of vibration[205] of the deflecting bar.

3. In measuring the deflection of a suspended auxiliary bar produced by the deflecting bar, at two different distances of the latter, in a south and north, or east and west direction from the magnetometer.

To this general view of the arrangements of the magnetical observatory, and of the apparatus to be placed therein, may be added the following remarks on the separate parts of both.

9.1 Remarks on the Separate Parts of the Magnetic Observatory, and of the Magnetic Instruments

1. *The room.* — Plates I and II are a perspective view and a ground-plan of the room. In the first the southern wall is supposed to be removed; in front, on the right, is seen, a, the foundation for the theodolite; b, the stand of the theodolite; c, the theodolite; d, the scale attached to the stand; e, the plumb-line suspended from the centre of the object glass. Near to it is stationed the clock, f, a line drawn from the theodolite telescope to the mark designated by the arrow on the opposite wall, would represent the magnetic meridian. Towards the centre the suspender of the magnetometer is fixed to the ceiling; from this is suspended the thread carrying the stirrup, in which is placed the magnet bar, to the anterior extremity of which the mirror is fastened vertically. The distance of the mirror from the telescope and its distance from the centre of the scale, (before which passes a plumb-line let fall from the theodolite telescope,) are, together, equal to the distance of the telescope from the mark.

2. *The theodolite.* — For observing the changes of declination, a telescope, having motion in a vertical plane, so that it may from time to time be directed either towards the mirror or towards the mark, is quite sufficient. This movement serves to ascertain and verify the stability of the telescope. For absolute measurements of declination a theodolite is employed instead of such a telescope. As the divisions of a scale divided into millimetres must not only be seen but even their subdivisions estimated, it is necessary that, at a distance of five metres of the scale and of the telescope from the mirror, the telescope should possess a magnifying power of at least thirty.

3. *The clock.* — All observations must be made accurately to time, for which purpose a clock which beats seconds must stand near the observer, with its face towards him. A chronometer may serve the purpose.

4. *The magnetometer.* — Besides a clock and a theodolite, which must be supposed present in all establishments where magnetic observations are to be executed in the most

[205][Note by AKTA:] In German: *Schwingungsdauer.*

perfect manner, the magnetometer consists of the following parts, which are necessary for measurements of declination: — the magnet bar, the stirrup with its torsion-circle, the suspender with screw and suspension thread, the mirror and mirror-holder, the torsion-bar, the scale, and the quieting bar; to which must be added, for measurements of intensity, the measuring scales, the deflecting bar, the weights, and the weight-holder. The magnet bar, in its connexion with the stirrup and the torsion-circle, (which again is connected by the suspension thread with the suspender,) and with the mirror and mirror-holder, is represented in Plate III, Figures 3 and 5.[206]

5. *The scale.* — Figure 10 gives a specimen of the scales hitherto employed,[207] which must be at least one metre in length. M. Rittmüller of Göttingen has lithographed such a scale, and has had it printed on white card-paper.

6. *The plumb-line at the object glass of the telescope.* — A fine wire of dark colour with a weight at its lower extremity, is fastened in such manner to the upper rim of the object glass, that it hangs correctly over its centre. In order to fix this wire, the small notches of the grooved frame of the object glass may be used; or a ring, constructed specially for this purpose, may be slid over the frame, having two slits diametrically opposite each other. The upper slit serves for the fastening of the wire, and the ring is so arranged that the wire passes freely through the lower. If we now view the image of the scale in the mirror through the telescope, we see at the same time the image of this wire projected on the white surface of the scale, and can thus find that point of the scale which lies in the vertical plane of the optical axis of the telescope. The spot where the prolonged plumb-line touches the ground is carefully marked, and serves as a means of testing the immobility of the theodolite stand.

7. *The mirror and mirror-holder.* — The mirror of the magnetometer must be perfectly plane, because otherwise, with a magnifying power of 30, the image of the scale would be indistinct. The plane mirrors from Utzschneider's optical manufactory in Munich have hitherto proved the best. The mirror should be somewhat broader than it is high, as, by the vibration of the magnet bar, the right and left side of the mirror alternately enters the field of the telescope. The best dimensions of the mirror are from 50 to 70 millimetres in height, and from 70 to 100 in breadth. In measuring the distance of the mirror from the scale and from the mark, the refraction of the rays of light at the anterior surface of the glass must be considered: that plane is the reflecting plane which is equidistant from the anterior and posterior surfaces of the mirror. The mirror is fixed to that end of the magnet bar which is turned towards the telescope, and must form with it so solid a system that no reciprocal disarrangement of either may be feared during the experiment, although the magnet bar be taken out of the stirrup and replaced in it in a reversed position. Moreover, the mirror must have such a position relatively to the bar, that the normal of the mirror shall be quite, or very nearly, parallel to the magnetic axis of the bar. The mirror holder represented at Figure 4 may serve both these purposes;[208] its frame is attached by screws to the bar. The frame-work supporting the mirror may be turned by screw motion round two rectangular axes, by which it may be brought into the required position.

8. *The suspender, elevating screw, and suspension thread.* — It is very advantageous to fix to the ceiling the thread, which is to carry the magnet bar, as by this it is sufficiently insulated from the floor and protected from all shaking, and because a proper length may

[206][Note by AKTA:] Plate III appears on page 101. Figure 3 appears separately on page 96 and Figure 5 on page 98.

[207][Note by AKTA:] See page 99.

[208][Note by AKTA:] See page 97.

in this manner be given to the thread. If, for the support of the magnetometer, we employ, not a metal wire, (the elasticity of which for an equal tenacity is almost ten times greater than that of one formed of silk fibres,) but a thread composed of parallel fibres of raw silk, it lengthens greatly, especially at first; and it is therefore requisite from time to time to raise it, so that the magnet bar and the mirror fixed to it may regain their original height. In raising the thread it is necessary that it should not be displaced in a lateral direction. A screw may be employed for this purpose, in the grooves of which the thread lies, and upon which it can be wound up still further, while the end of the screw works into a fixed nut. The groove in which the thread places itself, by the turning forwards of the screw, takes then, of itself, (from the advance of the whole screw) the place which the vertically suspended line had before occupied. The fixed nut, with solid rest, through which the screw pin passes freely, is let into a wooden slider which is mortised into a large plank fixed to the ceiling, and can slide therein in a direction parallel with the north or south wall of the room. If the position of the magnetic meridian should change in the lapse of time in any considerable degree, this slider will serve to retain the magnetometer in the meridian of the telescope. After such a sliding of the suspender on the ceiling, which need be performed but very seldom, it is necessary to place on the opposite wall a new mark, to which the telescope may be directed without departure from the meridian. The thread to which the magnet bar is suspended consists of 200 parallel fibres of raw silk, each of which would support thirty grammes without breaking. The weight which this thread has usually to sustain amounts to nearly 2000 grammes, to which, in the measurements of Intensity, two weights of 500 grammes are added when determining the moment of inertia of the magnetic bar. The thread, therefore, never carries more than half the weight with which it would break. It is about two metres long, and has a torsion force, the moment of which amounts, for small deviations, to about the 1000th part of the magnetic [torsion] force. The thread may be prepared by winding a single fibre twenty-five times round two glass tubes, distant from each other about four times the intended length of the thread; the two ends of the fibre are then tied firmly together, and the twenty five-fold skein,[209] thus formed, is stretched by drawing the two glass tubes further from each other. A small hook, carrying a weight, is then attached to the skein, midway between the two tubes, which are then raised and brought together, and the two loops are united in one. Thus a hundred-fold thread is prepared, which forms a loop at top and bottom, and which, being again brought together in a similar manner, forms the thread to which the magnet bar is suspended.

9. *The stirrup and torsion-circle.* — The force of torsion of the thread to which the magnet bar is suspended must not be entirely neglected in absolute measurements of declination and intensity, even though this thread be very long and fine. In order to measure the magnitude of this force, and to diminish its influence, so that the thread in the mean position of the magnet bar may be brought to its natural position when its moment of torsion is zero, it is necessary to be able to turn the thread, at one of its two extremities, round itself, in such manner that the angle of torsion may thereby be measured. In order to have the means of effecting this at hand, the apparatus for this purpose must be at the lower extremity of the thread; but, that the magnet bar may not be turned with it, the stirrup is composed of two parts, an alidade and a circle, which revolve only round a common vertical axis. The alidade supports the magnet bar, and is itself supported by the circle; the circle is provided with a pivot which passes through the alidade, and has, at its upper extremity, two hooks to receive the pin fixed to the thread. With this arrangement of the stirrup, it is

[209][Note by AKTA:] In German: *Ring.*

important that the alidade in which the magnet bar lies should rest on the rim of the circle; otherwise, the friction, taking place near the axis of rotation, would produce a displacement of the parts relatively to each other, in consequence of the impulse arising from the vibrating bar. Moreover, the stirrup is so constructed that the magnet bar fits in either on its broad or narrow side. This is done for the purpose of determining accurately, by observations of declination in any of the various positions of the magnet bar in the stirrup, the position of the mirror relatively to the magnetic axis of the bar.

10. *The box and the measuring scales.* — The box which protects the magnetometer from the influence of currents of air is constructed so as to afford ready access to the instrument within. It forms a cylinder of 800 millimetres in diameter, and 300 in height. The cylindrical form is given to it for this reason; in the measurement of intensity, in order to ascertain the moment of inertia, a wooden rod 700 millimetres in length is placed at right angles on the magnet bar of 600 millimetres in length, and this rod, to which weights are suspended, must find a place in the box along with the magnet bar, and must vibrate freely. In order to perform these experiments with convenience, it is also requisite that the box should admit of being entirely opened at the top, and of being tightly closed again, so that there should only remain an aperture at the top for the suspension thread, and one for the mirror at the side. The latter may be closed with a small wooden slider, to exclude air when not observing. The box is closed above by two semicircular lids, which must fit exactly, one of which is provided with a small aperture for the thread. This aperture is not situated in the centre of the circle formed by the two semicircular lids, but is so placed that the thread passing freely through it, the mirror of the magnet bar may hang close before the aperture in the side of the box. This arrangement is necessary, in order that a small aperture may suffice to allow the light to pass from the scale to the mirror, and from that to the telescope. Around the case are fixed the measuring scales on which may be placed a second magnet bar to the south or north, to the east or west of the magnetometer, at prescribed distances and in a prescribed position, deflecting the suspended bar from the magnetic meridian.

11. *The torsion-bar and deflecting bar.* — That the thread to which the magnet bar is suspended is without torsion in the mean position of the latter, is recognised thus: a brass bar of equal length and breadth, and of nearly equal weight, as the suspended magnet bar, having a small magnet inserted in it (in order somewhat to shorten the duration of the vibration due to the elasticity of the thread) is placed in the stirrup instead of the magnet bar. If the thread is without torsion, the magnetic axis of the small magnet will be in the same line as that of the larger bar was. In order to test this accurately, the auxiliary bar must, like the principal bar, be provided with a mirror and a mirror-holder. For measurements of intensity a second magnet bar of like dimensions to the principal bar is required, which may also be placed in the stirrup instead of the latter, in order to observe its vibrations, and to measure its moment of inertia. The same bar, however, must also serve as a deflecting bar, and for this purpose it is fitted into a small wooden case, which is bounded exteriorly by even surfaces and straight edges parallel to its magnetic axis, in order to give it its place quickly and accurately on the measuring scales.

12. *The weights and weight-holders.* — For measurements of intensity it is requisite that the deflecting bar may also be vibrated, and its moment of inertia thus deduced. For this purpose a thin wooden rod is placed across the vibrating magnet bar, and two equal weights are suspended, at various distances from each other, successively on both sides of the magnet bar. In order to mark the points of suspension, and to determine accurately their mutual distances, both weights, each of which amounts to 500 grammes, are provided

with a small capsule. The capsule is placed on a fine point, projecting from the wooden rod. There must be several such projecting points at 50 millimetres distance from one another, with the exception of the two central ones, which are situated at 100 millimetres from each other. These distances must be measured with microscopical accuracy.

13. *The quieting bar.* — In order to perform the observations promptly and accurately, it is of importance to be able to moderate at pleasure the vibrations of the magnet bar; for instance, when measuring the duration of vibrations, to make the commencing arc no greater than 2 or 3 degrees, and in observing changes of direction, to make the arc as small as possible, never allowing it to exceed 2 or 3 minutes. This end is attained with the quieting bar, in the use of which every observer must practise himself. It is a magnetic bar half the length and breadth, and four times lighter than the principal bar. When this bar is held by the observer behind the theodolite in a horizontal position, and at a right angle with the magnetic meridian, it will cause at this distance (about 5 meters), if it is strongly magnetized, a deviation of about one minute, westerly if its north pole is held easterly, and *vice versa.* This deviation becomes smaller in proportion as the bar is removed from the horizontal position, and disappears entirely with its approach to the vertical position. No inconvenience is therefore occasioned by such a bar standing by the wall or near the clock case (as in Plates I and II), till wanted. The use of the quieting bar in magnetic measurements is manifold; and it is important, in order to attain perfect and skilful facility in the performance of these experiments, to become accurately acquainted with its mode of operation. A separate article will therefore be allotted subsequently to the explanation of the rules and laws for its various uses and modes of action.

Finally, the building may be situated in the neighbourhood of other buildings without any injury to the observations. The magnetic observatory in Göttingen, for instance, could not, without causing many difficulties, be situated far from the astronomical observatory. The magnetometer is stationed about 60 metres westward of the astronomical observatory. At this distance moderate magnetic forces exercise so small an influence on the magnetometer, that it has been found unobjectionable to erect in a room of the astronomical observatory an auxiliary magnetic apparatus, which is of very essential service in absolute measurements.

More accurate directions will be given subsequently for determining the influence of a distant magnet, according to its force and position relatively to the magnetometer; and will especially serve this purpose, that when several magnetical apparatus (for instance, a principal magnetometer, an auxiliary magnetometer, and an inclinatorium) are to be fixed in neighbouring buildings, a positive conviction may be acquired, that their influence on each other is harmless, or, if this should not be the case, that their effect may be reduced to calculation.

9.2 Explanation of Plate III

In this plate the several parts of the magnetometer are represented, with the exception of the clock, theodolite, measuring scales, the box, the torsion and quieting bar, which partly require no particular representation, and in part have been already shown on a smaller scale, in Plates I and II. On the other hand, the arrangement of the suspender with the elevating screw, the stirrup with the torsion-circle, the mirror-holder, with its corrections, the weights and the weight-holder, stand in need of a more accurate representation, which is given from various sides in this plate, on a scale of half their actual magnitude. The stirrup, the torsion-circle, and the magnet bar in its place, have been represented in three different positions

— from the west, from the south, and from above; the mirror-holder, and the suspender, with the elevating screw, have been figured from two sides — from the west, and from the south. In the south view of the stirrup, with the torsion-circle and the magnet bar in its place, is shown the manner in which the weight-carrier may be placed on the magnet bar in a west and east direction, and the two weights, each of half a kilogramme, suspended to the points with which it is furnished, for the purpose of determining, in absolute measurements of intensity, the moment of inertia of the vibrating portion of the magnetometer. To spare room on the plate, the two views of the bearer, with the elevating screw, have been placed in the upper series, close to one another, but this has prevented the bringing of the two into the correct position relatively to the vibrating portion of the magnetometer suspended from them. It is, however, easily seen how the view of the suspender, with the elevating screw in Figure 1,[210] is connected with that of the stirrup, torsion-circle, magnet bar, and mirror holder in Figure 3 if we attend to the commencement indicated in Figure 1 and the termination indicated in Figure 3 of the vertical line connecting them. These two Figures represent the main parts of the magnetometer in a westerly view. In the same manner Figure 2 and Figure 6 are connected,[211] and represent the instrument as observed from the southern position. In Figure 6 the mirror-holder has been taken off from the southern extremity of the magnet bar, so that it might not conceal the stirrup situated behind it, and is represented by itself in Figure 4. In the westerly view, Figure 3, the small notch in the stirrup into which the weight-carrier fits, is merely indicated; while in the southern view, Figure 6, it is shown as fitted into the notch, and placed on the magnet bar, and the two half kilogrammes it is to bear are suspended from its points.

Figure 1 presents a view of the suspender, with the screw and suspension thread, from the west.

Fig. 1.

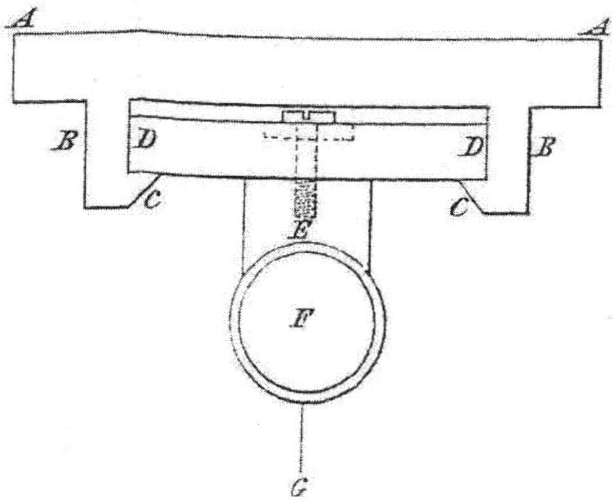

AA is a board fixed to the ceiling; *BB* two parallel wooden rods glued to it, between which a slider, *DD*, may be moved from east to west; it is supported by two projecting parts, *CC*; the brass nut, *E*, through which the elevating screw passes in a direction from east to west, is fixed with screws to the slider; *F* is the screw head at the western extremity, which in this Figure hides the screw; *G* is the suspension thread attached to the screw.

[210][Note by AKTA:] See page 95.
[211][Note by AKTA:] See pages 96 and 98.

Figure 2 represents a view from the south, of the same suspender, with the screw and thread.

Fig. 2.

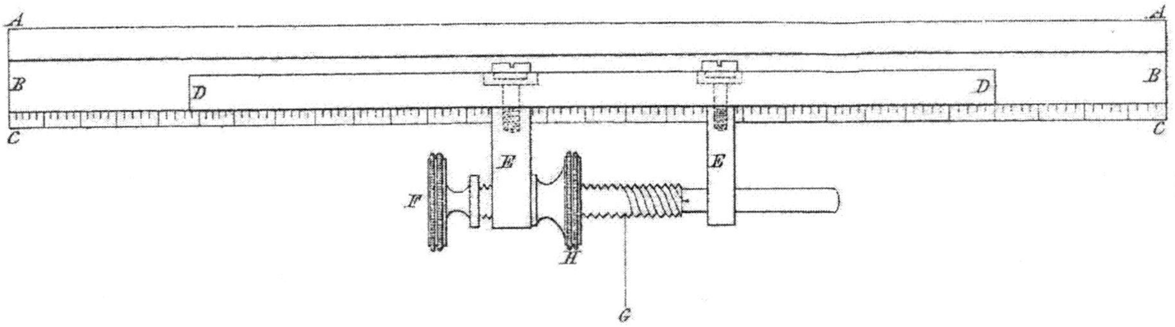

AA here, is the longitudinal section of the board fixed to the ceiling; BB is the rod glued to this board on the north side; CC the support of the slider; it is furnished at the edge with a scale, for the adjustment of the slider; DD the longitudinal view of the slider, to which the copper nuts E and E' are fixed with screws. Through these nuts passes the elevating screw, the head of which is represented by F. This screw passes through the nut E, and is kept in its place by the nut H. Near to the second nut E' the screw changes into a smooth cylinder which passes through a smooth aperture of the nut E'. At the end of the thread of the screw the suspension thread G is fastened, and lies in the grooves, in which it continues to the centre, and there falls perpendicularly, bearing at its lower end the stirrup of the magnetometer. When the thread is to be raised, the nut H is loosened, and the screw turned by the screw head F into the required position.

Figure 3 presents a view, from the west, of the vibrating portion of the magnetometer.

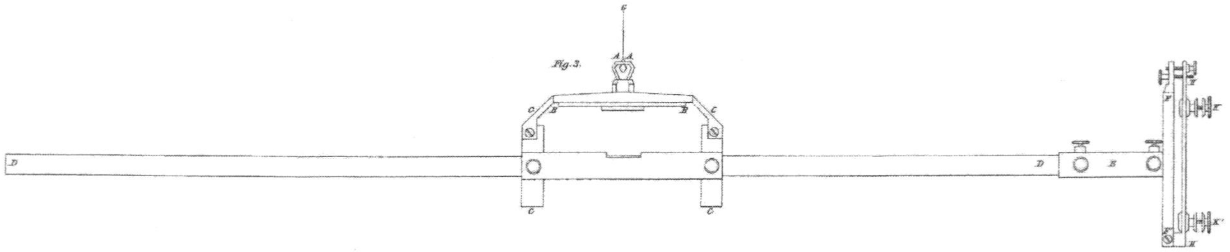

It consists of two eyes, AA, of which the posterior is concealed in this Figure by the anterior. The lower end of the thread G is fastened to a pin fixed under them. To this part of the magnetometer belongs also the torsion-circle BB, upon which the stirrup $CCCC$ rests; the magnet bar DD, and the mirror-holder E, with two frames FF, HH, and the clamps KK, serving to receive the mirror. With the exception of the magnet bar, which alone weighs 1700 grammes, and of the mirror, which must be of such thickness that it may not bend, all the other parts are constructed of thin brass, so as to increase the moment of inertia of the magnetometer as little as possible. The thread supporting the stirrup is not fastened immediately to it, but to a pin which fits below the staples AA, so that without unfastening it may be disengaged from the stirrup. The pin is provided with two small points, at a distance of about 40 millimetres from each other, which fit into two depressions on the staples AA. The torsion circle BB is furnished with a vertical pivot, the upper end of which supports the staples AA, and is surrounded by the rotating stirrup. The stirrup itself rests upon the periphery of the torsion circle, but is prevented from turning by its friction against it. At the end of the magnet bar DD is observed the mirror-holder, which at E forms a sheath incasing the magnet bar, to which it can be tightly fastened by screws. To this sheath is attached a frame FF' turning round a vertical axis. Small pressing and tightening screws, which serve for placing and fixing this frame, are behind it in this view, and therefore are not seen. With this first frame EF', turning round a vertical axis, is connected a second frame HH, turning round a horizontal axis at F', which can be adjusted to the first by means of the screws shown above. The clamps which are to receive the mirror are attached to this second frame. Three such clamps exist; but in this Figure only two, K and K', are visible, while the third is covered by the second at K'.

Figure 4 serves to give a more distinct view of all the parts of the mirror-holder, which, here seen from the south, are severally better seen than in the foregoing view from the west.

Fig. 4.

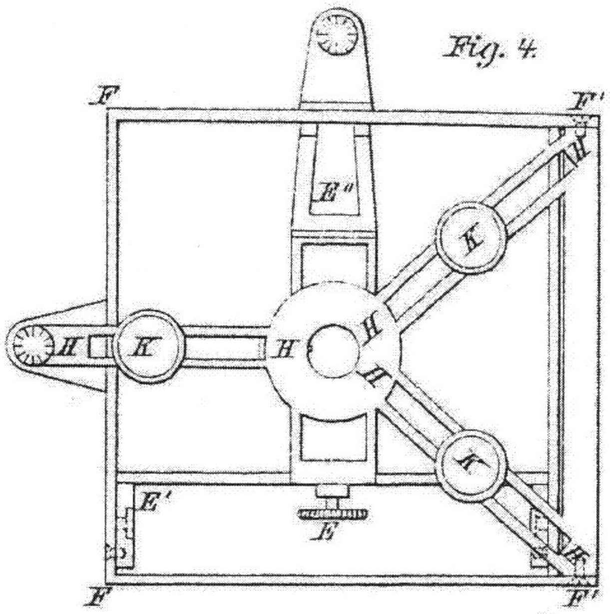

Each part is designated by the same letter. The rectangle seen between E and E'' is the transverse section of the sheath inclosing the magnet bar, to which it is firmly screwed. This case has on one of its sides two projections, $E'E'$, which form the vertical (horizontal in the Figure) axis of the frame $FFF'F'$. Opposite, near E'', is a third projection, against which the screws act, which serve for placing and holding fast this first frame. A horizontal (in our Figure vertical) axis is attached to this first frame at $F'F'$, around which the second frame $HHHH$ can revolve. Opposite to this axis both frames have small projections, whose relative distance can be adjusted by pressing and tightening screws. Three small incisions are shown, HH, HH, HH, into which three small sliders can be inserted and fastened. This arrangement serves the purpose of adjusting the space necessary for the reception of the mirror. These three small sliders terminate at their southern extremity, in three small vertical circular surfaces, on which the edges of the mirror are placed; while the head of a screw, whose grooves fit into the sliders beneath the edge of the mirror, press on its front surface. In this Figure the sliders themselves are not seen, but merely the heads of the three screws, which fit into and conceal them.

After these explanations of the first Figures, a few short remarks respecting the others will suffice.

Figure 5. In this view of the stirrup, torsion-circle, magnet bar, and mirror-holder, seen from above, the torsion circle is more distinctly presented to view, as also the form of the stirrup.

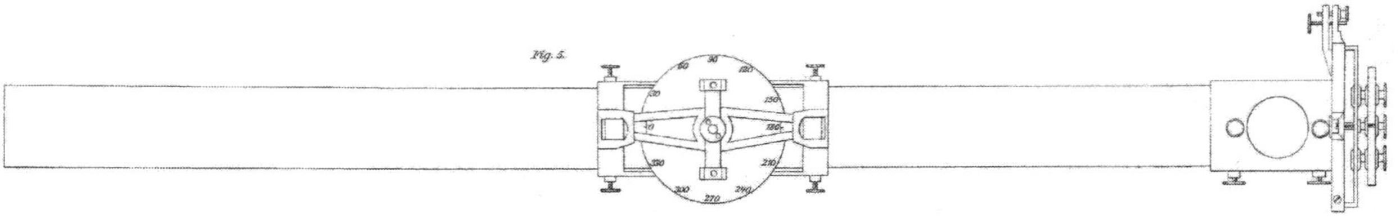

In the centre of the circle is also visible the end of the pivot passing through the alidade, and the double staple attached to it, with its two pivot holes. The brass pin, whose points fit into these holes, is removed, for the sake of perspicuity. In this Figure, moreover, is seen how the mirror is fastened to the mirror holder.

Figure 6. In this Figure, which has often been referred to previously, is chiefly seen in what manner the points of the pin, to which the suspension thread is fastened, fit into the holes of the staples, which latter are connected by a centre-piece provided with a square aperture in its own centre, into which the 4-sided pin of the torsion circle is inserted, and held fast by a screw.

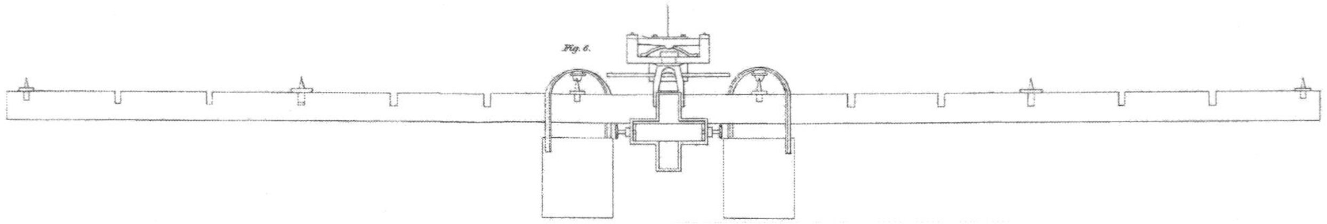

Since the stirrup, together with the magnet bar, must be raised when the latter has to be inverted for the purpose of finding its magnetic axis, the pin to which the thread is fastened would then fall out, but for a small spring beneath, which is visible in this Figure, and which then retains the pin in its position. The wooden rod, above 700 millimetres in length, which in this Figure is laid across the centre of the magnet bar, and serves for the support of two half-kilogrammes which are to increase the moment of inertia of the magnet bar, is furnished with 6 points, on which the two weights can be placed at different distances. The two central points are at a distance of 100, the next two at a distance of 400, and the extreme points at a distance of 700 millimetres from each other. The first and last are fixed; the two intermediate ones can be taken out and placed in other notches, situated at distances of from 50 to 50 millimetres asunder. The distances of all these points must be measured with microscopical accuracy.

Figures 7, 8, and 9 represent the pin to which the thread is fastened, seen from one side, from above, and from below.

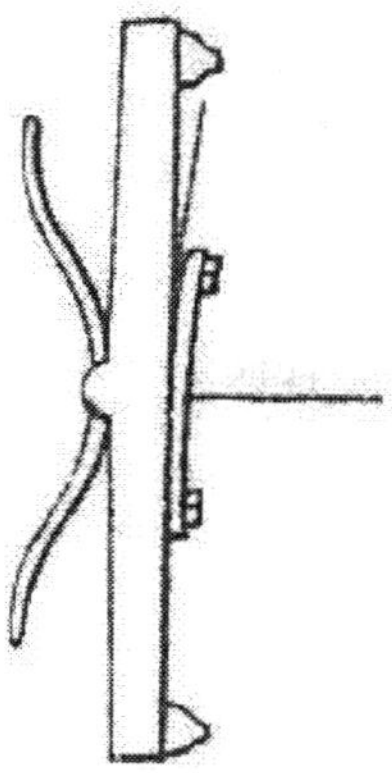

Fig. 7. Fig. 8. Fig. 9.

The first view exhibits the two points with which this pin fits into the holes of the staples of the torsion circle, as also the spring which retains it when the stirrup is raised, and the thread loosened. The second view shows the narrow, round aperture through which the thread passes and is held together. The third view exhibits an oval aperture, which is bisected by a round transverse pin. The thread is wound round this latter, and drawn tight, after having been longitudinally drawn through a loop formed by its inferior extremity.

Figure 10, gives a representation of the scale which is fixed below the theodolite, and the reflected image of which is observed with the theodolite telescope.

Fig. 10.

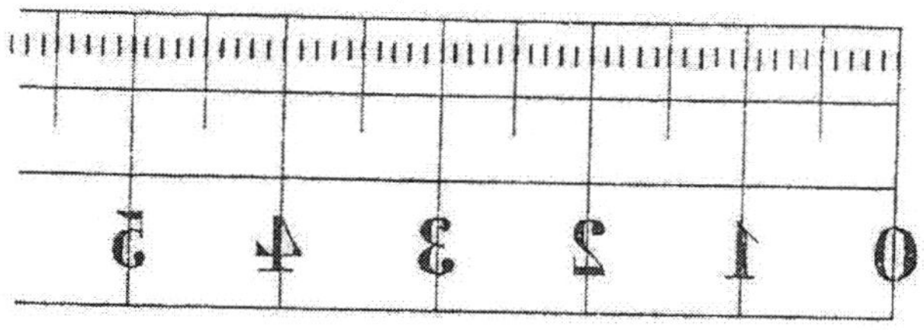

By employing an astronomical telescope (which, with a similar object-glass, is preferable, for clearness and definition, to the terrestrial telescope) the scale is inverted, so that the figures stand above the divisions, while, in our Figure, they are situated beneath them.

9.3 Expense of Building and Furnishing a Magnetic Observatory

The expenses consist in the cost of the *building* and the *instruments*. That of the building is not everywhere the same; at Göttingen, it amounted to 798 dollars,[212] Prussian currency. A part of the costs were occasioned by the exclusion of iron in the nails, locks, hinges, and fastenings of all kinds, all of which are of copper.

The costs of the instruments, as supplied by Meyerstein of Göttingen, who has hitherto made the greatest number of such instruments, are as follows:

1. An 8-inch theodolite: 150 Dollars.
2. A seconds clock: —

[212][Note by WF:] The Prussian current dollar is equal to three shillings.

3. A stand for the theodolite: 7 Dollars.

4. A scale, with frame: 1 Dollar.

5. The illuminating apparatus: 11 Dollars.

6. The suspender, with slider and screw: 8 Dollars.

7. The stirrup, with torsion circle: 15 Dollars.

8. A 4-lb. principal bar, with its case; a 4-lb. auxiliary bar, and a 1-lb. quieting bar: 7 Dollars.

9. A brass torsion bar, with magnets inlaid: 9 Dollars.

10. Two mirror-holders, with adjustments and mirrors: 43 Dollars.

11. A weight-holder, with two half kilogrammes with hooks: 7 Dollars.

12. A case with glass lid: 16 Dollars.

13. Three measuring scales, 6 metres long: 4 Dollars.

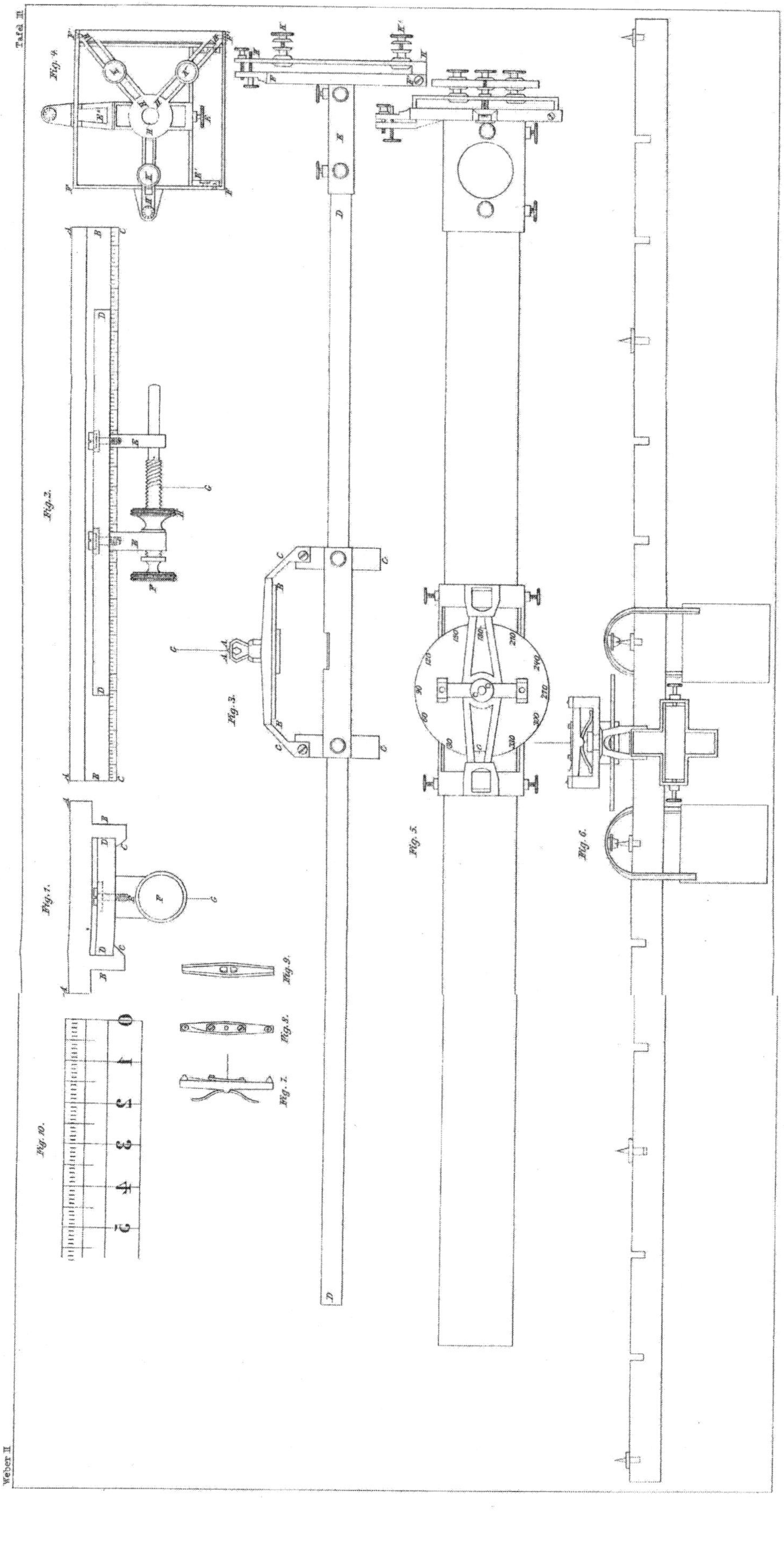
Tafel II
Weber II
Fig. 1.
Fig. 2.
Fig. 3.
Fig. 4.
Fig. 5.
Fig. 6.
Fig. 7.
Fig. 8.
Fig. 9.
Fig. 10.

Chapter 10

[Weber, 1837b] Description of a Small Portable Apparatus for Measuring the Absolute Intensity of Terrestrial Magnetism

Wilhelm Weber[213,214]

Among the numerous applications of the magnetometer,[215] the most important is that of measuring the absolute intensity of the earth's magnetic force, as described in the memoir entitled, *Intensitas vis magnetica terrestris ad mensuram absolutam revocata; Auctore Carolo Friderico Gauss: Göttingen, 1833.*[216,217] Frequent mention will be made in the course of this work of this application of the magnetometer, which enables us to compare numerically with one another the results of experiments made in the most distant parts of the globe, at different epochs, and with apparatus not previously compared. Everything necessary to be known for these experiments, as well as everything that may serve to facilitate them, will be communicated from time to time. Results of such absolute measurements will also be noticed, and their value shown in establishing, on a scientific basis, the science of galvanism.

These important absolute measurements can be performed with the accuracy they deserve only with the magnetometer, and, indeed, only in a completely furnished observatory. Few such observatories, however, exist at present, and few philosophers, therefore, have these means at their disposal; while there are many who take an interest in the results, and desire to be enabled to form such an opinion concerning them as can hardly be satisfactorily obtained without actually taking part in the observations and calculations, even though less minute and accurate. The simple means which it is the object of this chapter to describe may

213[Web37b] with English translation in [Web41a] and French translation in [Web38b].

214The Notes by Wilhelm Weber are represented by [Note by WW:]; the Notes by E. Riecke, the editor of the second volume of Weber's *Werke*, are represented by [Note by ER:]; the Notes by Carl Friedrich Gauss (quoted by Weber in this article) are represented by [Note by CFG]; while the Notes by A. K. T. Assis are represented by [Note by AKTA:].

215[Note by AKTA:] A detailed description of the magnetometer can be found in [Web37a] with English translations in [Web41f], [GW66] and [GW39a]; and French translation in [Web38c].

216[Note by ER:] Gauss' *Werke*, Vol. V, p. 79.

217[Note by AKTA:] See Chapters 5, 6 and 7.

be procured by every person. The description and mode of employing them are with the more propriety given here, because these pages are intended not merely for the limited number of those who participate in the simultaneous observations, but for all who are engaged in investigating the laws of magnetic phenomena.

Those less delicate instruments which were employed for magnetic measurements before the invention of the magnetometer, may not only be used for the same purposes as formerly, but may also be applied to those absolute measurements of intensity which owe their origin to the invention of the magnetometer. It is true that these instruments are far from affording such accurate results as the magnetometer; but the results they give are more easily obtained. On this account they have not lost all their value by the later invention; they may still be usefully employed, though in a more limited sphere. Wherever, from want of means or time, or from any other circumstances, magnetometers cannot be employed, these instruments may still be used with advantage. This will be the case most frequently in voyages and journeys to remote parts of the world. It is true that magnetometers may be carried on journeys, as was done by M. von Waltershausen and Dr. Listing in their Italian tour;[218] but this is only possible for those who are highly favoured by external circumstances; and it is therefore not to be expected that many will follow this praiseworthy example. If, therefore, we wish to collect observations from the whole surface of the earth, we must be content with such as are not made with magnetometers; and it is important to extend the application of portable instruments to the absolute measurement of the intensity, which has been hitherto performed with the magnetometer only. The difference in respect to accuracy between the absolute measures with such instruments and those made with the magnetometer, is nearly the same as between measurements of declination with the two kinds of instruments. A skilful hand will be able to obtain useful results even with the smaller apparatus; and it appears desirable, therefore, that it should be extensively employed.

We shall consider successively,

1. The parts of the small apparatus.
2. The observations to be made with it.
3. The application of the observations.
4. The calculations required.
5. The result of the calculation.
6. The advantages, in point of accuracy, of the dimensions adopted in the apparatus.

10.1 The Parts of the Small Apparatus

In addition to a clock or chronometer, this apparatus consists of three parts:

1. A small compass needle.
2. A small magnet bar, which may be suspended to a silk thread, and vibrated.
3. A measuring scale 1 metre in length.

The needle of the compass from which the present description is taken was 60 millimetres in length, and the arc was divided to whole degrees only. In order that so small a compass should lead to useful results, it is necessary that the observer should be able to estimate with certainty the 10th part of a degree.[219] The needle may be somewhat larger; but the

[218][Note by AKTA:] Wolfgang Sartorius von Waltershausen (1809-1876) and Johann Benedict Listing (1808-1882).

[219][Note by WW:] This estimation, which under other circumstances is easy to accomplish, presents in this

reasons, which render it advisable that it should never exceed 100 millimetres, will be given at the conclusion of this chapter.

The small magnet bar was 101 millimetres in length, $17\frac{1}{2}$ in breadth, and 142 grammes in weight; it may be vibrated by suspending it to a silk thread bound crossways round the middle of the bar. It is advantageous that it should be made an exact parallelopiped, in order that, its weight and dimensions being known, its moment of inertia may be calculated. It may also be provided at the middle with a small hole, through which a sewing needle may be passed, in which case it is merely necessary to draw the suspension thread through the eye of the needle; it is better to make the small bar precisely 100 millimetres in length.

The breadth of the measuring scale must be such as to allow of the compass being placed on its centre; this scale need only be divided to 50 millimetres.

This simple apparatus is sufficient for the absolute measure of the magnetic intensity. It is furnished by M. Meyerstein,[220] of Göttingen, for 9 dollars and a half, (of course, exclusive of the time-piece); so that this mode of measuring the intensity requires less expense than any other magnetic determination. It is also very portable and convenient for travellers. The apparatus is to be placed on a table in the middle of a room, avoiding all iron in the neighbourhood; large iron rails, even at some distance, must be carefully avoided. Arrangements may also easily be made for employing it in the open air.

10.2 Observations to be Made with This Apparatus

These are of two kinds: 1. The experiments of deflection. 2. The experiments of vibration.

10.2.1 The Experiments of Deflection

The measuring scale is placed horizontally, and at right angles to the magnetic meridian, with its zero point towards the east, and the needle in the centre. The small magnet bar is to be placed successively as follows:

1. With its north end to the east, on the zero point of the scale: if the length of the small magnet bar is 100 millimetres, its centre will then be over the division 50^{mm} of the scale. The needle will be deflected towards the east, and its position, u_0, is observed.

2. The south end is then substituted for the north end. The needle is deflected to the west, and its position, u'_0, observed.

3. The north end of the magnet bar is placed towards the east, on the division 100 millimetres. The needle is deflected easterly, and its position, u_1, observed.

4. The bar is again reversed, end for end. The needle is deflected westerly, and its position, u'_1, observed.

5. The north end of the magnet bar is placed towards the east, upon the division 150 millimetres. The needle is deflected easterly, and its position, u_2, observed.

6. The bar is again reversed, and the needle deflected westerly, and its position, u'_2, observed.

---

case some difficulty, arising from the point [tip] of the magnetic needle being usually at a little distance from the divided arc: to get over this difficulty the following method is adopted: a mirror is laid horizontally on the table before the needle, and the eye, before it reads off the position of the needle, must be so placed that the prolongation of the needle would bisect the reflected image of the eye.

[220][Note by AKTA:] See footnote 200 on page 85.

7. The north end of the magnet bar is placed towards the east, upon the division 750 millimetres. The needle is deflected easterly, and its position, u_2'', observed.

8. The bar is reversed. The needle is deflected westerly, and its position, u_2''', observed.

9. The north end of the magnet is placed towards the east, on the division 800 millimetres. The needle is deflected easterly, and its position, u_1'', observed.

10. The bar is reversed. The needle is deflected westerly, and its position, u_1''', observed.

11. The north end of the magnet bar is placed towards the east, on the division 900 millimetres. The needle is deflected easterly, and its position, u_0'', observed.

12. The bar is reversed. The needle is deflected westerly, and its position, u_0''', observed.

These twelve observations may all be completed in half an hour.

10.2.2 Experiments of Vibration

The small magnet bar is to be suspended horizontally by a silk thread, to be set in vibration, and its time of vibration observed in the usual manner, which needs no further description here. The time of vibration may be determined by these experiments, with sufficient precision, in a quarter of an hour.

Taking together all the observations which are necessary for a complete measure of the absolute intensity, and allowing a quarter of an hour for arranging the apparatus and suspending the magnet bar, the experimental part of the determination can be completed in one hour. The observer may give his determination greater certainty and accuracy by repetition.

The following observations made with this instrument at Göttingen are given as an example.

Göttingen, January 18, 1837.

1. Experiments of deflection.

1.	$u_0 - u_0' = 23°9'$
2.	$u_1 - u_1' = 47°42'$
3.	$u_2 - u_2' = 71°48'$
4.	$u_2'' - u_2''' = 69°21'$
5.	$u_1'' - u_1''' = 46°12'$
6.	$u_0'' - u_0''' = 22°27'$

In these experiments, the distance, R, of the centre of the small magnet bar from the centre of the compass, was successively,

1.	$R_0 = 450\ mm$
2.	$R_1 = 350\ mm$
3.	$R_2 = 300\ mm$
4.	$R_2 = 300\ mm$
5.	$R_1 = 350\ mm$
6.	$R_0 = 450\ mm$

2. Experiments of Vibration.

No.	Clock Time	Number of Vibrations	Their Interval.
0.	$0'\ 3.25''$		
1.	$9,90''$	1	$6.65''$
2.	$16.65''$	2	$13.40''$
3.	$23.35''$	3	$20.10''$
4.	$30.00''$	4	$26.75''$
5.	$36.65''$	5	$33.40''$
6.	$43.30''$	6	$40.05''$
7.	$50.00''$	7	$46.75''$
8.	$56.70''$	8	$53.45''$
9.	$1'\ 3.30''$	9	$60.05''$
10.	$9.80''$	10	$66.55''$
11.	$16.55''$	11	$73.30''$
12.	$23.30''$	12	$80.05''$
13.	$29.90''$	13	$86.65''$
14.	$36.65''$	14	$93.40''$
15.	$43.15''$	15	$99.90''$
16.	$49.80''$	16	$106.55''$
17.	$56.65''$	17	$113.40''$
18.	$2'\ 3.25''$	18	$120.00''$
19.	$9.95''$	19	$126.70''$
20.	$16.70''$	20	$133.45''$
21.	$23.35''$	21	$140.10''$
22.	$30.00''$	22	$146.75''$
		Sum 253	$1687.40''$

Consequently the time t of one vibration:

$$t = 6.67''\ .$$

10.3 Application of the Observations

A general and intelligible view of the application of these observations, without entering into theoretical considerations, will be best given by extracting certain passages from a memoir in Schumacher's *Jahrbuch* for 1836, entitled, "*Ueber Erdmagmetismus und Magnetometer*;"[221,222] and adding the mathematical expressions of the laws there given verbally.

> "The square of the number of vibrations made by a magnetic needle in a given time, is a measure of the intensity of the earth's magnetism which depends on the needle employed. The individual properties of the needle have a two-fold influence: — first, by the greater or less magnetic force which it possesses; and secondly, by the effect of its form and weight on the time of vibration. The elimination of the latter effect presents no difficulty. The influence of the earth's magnetism on the magnetism of the needle produces a force or moment of rotation when the needle is not in the magnetic meridian: this moment of rotation is greater, the more the needle deviates

[221][Note by ER:] Gauss' *Werke*, Vol. V, p. 315.
[222][Note by AKTA:] [Gau36b].

from the magnetic meridian; and is greatest when the needle is at right angles to that meridian. This maximum of effect is always to be understood when the moment of rotation simply is spoken of; it may be represented by a given weight acting on a lever of given length, and consequently by a number, if the weights and lengths are expressed in numbers, according to arbitrary units. Now this moment of rotation and the time of vibration are very simply connected, by means of an intermediate quantity, dependent on the figure and weight of the needle, called its moment of inertia, and which may be calculated according to known rules. If the needle is not a perfectly regular body, or if it carries any appendage when in vibration, other means are required for the determination of its moment of inertia, the description of which would lead us too far; suffice it to say, that it is always possible. The moment of inertia then being known, the moment of rotation produced by the earth's magnetism on the magnetism of the needle, may be concluded from the observed time of the vibration of the needle."

If we designate by the letter C, the moment of inertia, after it has been multiplied by π^2, *i.e.*, 9.8696... and divided by twice the height of the fall of a heavy body in the unit of time, we may conclude from C, and from the observed time of vibration t,[223] the greatest moment of rotation caused by the earth; it is

$$= \frac{C}{t^2} \ .$$

"It is possible to determine the moment of rotation by direct experiment, without observing the time of vibration. An apparatus, expressly adapted to this purpose, has been recently placed in the Göttingen Astronomical Observatory, and is susceptible of great accuracy; but for the present purpose it is unnecessary to dwell on this point.

"The moment of rotation, produced by the earth's magnetism on a given needle, offers a new way of measuring the force of the earth's magnetism, or, to speak more accurately, a new form of the previous mode of measurement, over which it has this advantage, that one portion of the individuality of the needle is thereby removed. The measurement is still dependent on the remaining peculiarity of the needle, namely, its own magnetism; and as soon as we can reduce *this* to an absolute measure, the force of terrestrial magnetism itself may also be reduced to an absolute measure; for we have only to divide the number which expresses the moment of rotation by the number which measures the magnetism of the needle. In fact, one unit of measure of the earth's magnetism is attributed to a force that is thought to be similar to it, whose action on the unit of magnetism of the needle consists in a moment of rotation, measured by the force which the unit of weight exerts on a lever of the unit of length."[224]

[223][Note by AKTA:] In German: *Schwingunsdauer*. This expression can also be translated as the "period of oscillation". See also footnote 176 on page 58.

[224][Note by AKTA:] The last sentence reads in the original as follows:

"In der That ist dann der Abmessung des Erdmagnetismus als Einheit eine solche diesem ähnlich gedachte Kraft unterlegt, deren Wirkung auf eine Einheit des Nadel-Magnetismus in einem Drehungsmoment besteht, welches durch den Druck der Gewichtseinheit auf einen hebelarm von der Länge der Raumeinheit gemessen wird."

This sentence was translated by W. Francis as follows, [Web41a, p. 72]:

If, therefore, T signifies the terrestrial magnetism, and M the magnetism of the needle, or of the vibrating bar,[225]

$$T = \frac{C}{t^2 \cdot M} \qquad (I.)$$

"We might be inclined to suppose that the weight which a magnetic needle can carry would afford a standard by which the force of magnetism developed in the needle might be estimated; but a closer examination will show that this method is quite unavailing for our purpose. The determination itself is incapable of much precision; for repeated experiments give very different results; but there is a still more important objection: the capability of sustaining weight has no necessary connexion with the magnitude of the development of magnetism in the needle, in the sense in which it must here be understood. The moment of rotation is due to the magnetism of all the parts of the needle, upon which the terrestrial magnetism acts equally, and in parallel directions. The sustaining power, on the contrary, is chiefly due to the magnetism situated in the ends nearest to the weight, which, moreover, is modified every moment by the reciprocal action of the magnet-bar and the suspended iron.

"A magnetic needle, at a given place, acts on every point of space, in an amount and direction determined by its distance and position. In the immediate neighbourhood its action is strong, but very unequal on different parts; at great distances the action is weak, but almost uniform in strength and direction within a moderate space. The greater the distance, the nearer the law of the force approaches to a rule, which is very simple, and is completely given by theory: we may limit ourselves here to the consideration of a single case, which is sufficient for our purpose.

"Let NS be a fixed magnet in a horizontal position; it is required to find its influence on a second needle, ns, suspended to a thread; the relative position of the two needles being shown in the annexed Figure:

"The action of the first needle upon the second will consist in imparting to it a tendency to turn in the direction indicated by the arrows, the letters N, n designating the North poles, and S, s the South poles. The moment of rotation is expressed by a number, exactly in the same way that the action of terrestrial magnetism on a

"In fact, the earth's magnetism is measured by a force equal to itself, whose action on the unit of magnetism of the needle consists in a moment of rotation, measured by the force which the unit of weight exerts on a lever of the unit of length."

I replaced this translation by the one appearing in the text.

A needle with one unit of magnetism has one unit of magnetic moment.

[225][Note by AKTA:] That is, M is the magnetic moment of the needle.

needle vibrating freely has been indicated above. The magnitude of the moment of rotation depends, however, on the distance, and on the magnetic force in *both* needles.[226] Thus for example (supposing the distances to be sufficiently great), at equal distances it would be augmented six-fold, by doubling the magnetism of one of the needles and trebling that of the other.

"The effect depends on the distance in such manner, that, at twice the distance the effect will be $\frac{1}{8}$th, and at three times the distance $\frac{1}{27}$th of the effect produced at the simple distance; bearing in mind, however, that this law is correct only for very great distances, and cannot be extended to small ones. As all distances, when referred to a selected unit, can be expressed by numbers, this law may be expressed thus: 'the moment of rotation, multiplied by the cube of the distance, is constant for very great distances.' This product may be termed with propriety the moment of rotation *reduced to the unit of distance*; remembering that, according to the remark above made, the actual moment of rotation at the unit of distance, when the distance is small, may differ considerably from the reduced moment. This, however, does not prevent us from employing the reduced moment of rotation as a measure of the magnetism of the needles, *and from considering as unity, the magnetism of that needle, which imparts to another needle, (of equal magnetism, and in the given position) a reduced moment of rotation equal to the effect of the unit of weight on the arm of a lever whose length is the unit of distance.*"[227]

If we represent, according to this established unity, the magnetism of the needle by m, that of the bar by M, the distance (supposed considerable) between them by R, and the moment of rotation exerted by the bar on the needle by f, the reduced moment of rotation is

$$mM = fR^3 .$$

The position of the bar relatively to the needle, assumed in this case, did not in fact exist in the Göttingen experiments, but a different position represented in the annexed figure.

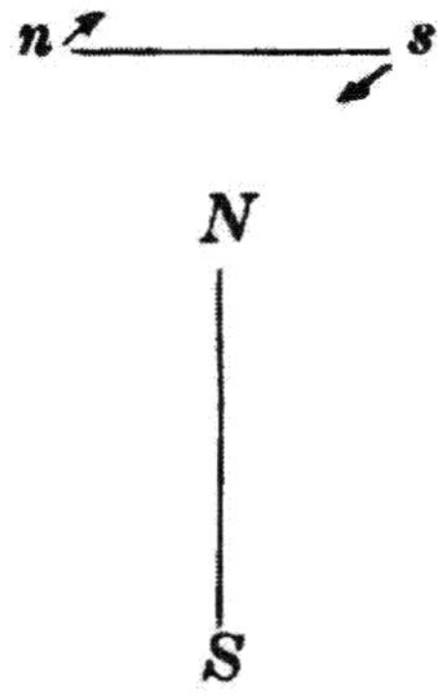

However, the same thing is true of the two positions, with this single difference, that f has a different value, which we shall designate by F. In the Memoir "*Intensitas*," etc.,[228] it

[226][Note by AKTA:] In German: *von der Stärke des Magnetismus in beiden Nadeln.* That is, the torque depends not only on the distance between both needles, but also on their magnetic moments, or on the strength of magnetism in both needles.

[227][Note by AKTA:] A needle with one unit of magnetism has one unit of magnetic moment.

[228][Note by AKTA:] See Chapters 5, 6 and 7.

is proved that

$$F = 2f \ ,$$

so that,

$$mM = \frac{FR^3}{2} \qquad (II.)$$

It is to this second case that the formulae, hereafter to be mentioned, will refer, as applicable to the Göttingen observations.

> "In this way therefore we have a complete and precise idea of the measure of the magnetic force of a magnetized needle. A needle of twofold power will impart to one equally magnetized a reduced moment of rotation = 4; and generally, when we know the number for the reduced moment of rotation which a needle imparts to another needle equally magnetized, we have the absolute measure of the power of magnetism in each needle; it being the square root of that number.
>
> "There only remains then, in order to be able to reduce the force of terrestrial magnetism to absolute measures, to give some method by which the moment of rotation which a needle produces in a similar one at considerable distance, (and in the position represented in the Figure) may be determined with precision. A great difficulty might at first appear, from a circumstance purposely omitted in what has been already said, viz. the impossibility of observing this very weak action of the needle NS upon the needle ns, (which we will for a time suppose to be magnetized exactly as strongly as NS); since it cannot be withdrawn from the omnipresent, and much more powerful action of the earth's magnetism. But this very circumstance affords the means of an easy solution. Let us suppose that in our Figure the straight line, from the centre of the magnet NS, through the needle ns, coincides with the magnetic meridian; in this position the terrestrial magnetic force will not act at all on the needle ns. As soon, however, as the moment of rotation which NS exerts on ns begins to act, ns will be deflected from its original position, and set in motion; but the more it deviates on account of this movement from its first position, the more strongly does the earth's magnetism tend to bring it back to its former position. The needle consequently performs vibrations about a line, which is no longer in the direction of the magnetic meridian itself, but is more or less inclined to it. This line is the position of equilibrium of the needle ns, which it assumes when the vibrations have ceased. This direction is evidently that of the resultant of the two forces, viz. the earth's magnetism, and the magnetism of the needle NS. According to the well-known laws of statics, the proportion of the strength of these forces, which is also the proportion of the moments of rotation produced by them, may consequently be determined from the angle of deviation, i.e. from the difference between the two positions of repose of ns, when it is subjected to the action of both the forces; and when NS is removed.
>
> "Here then arises another important remark; namely, that the angle of deviation of the needle ns is quite independent of its magnetism; as any increase in that respect evidently causes both moments of rotation to increase in the same proportion. We are thus freed from the necessity of fulfilling the difficult condition of equality in the magnetism of the two needles."

If we represent the deflection by v, — the greatest moment of rotation exerted by the earth on the needle (according to the measure fixed for the terrestrial magnetism) by mT, — and by F, the moment of rotation exerted by the magnetism of the bar (M) on the magnetism of the needle (m) at the distance R; the forces exerted by the earth, and by the bar, on the needle, are to each other in the proportion of the cosine to the sine of the deflection v; and the moments of rotation, mT and F being also in the same relation to each other,

$$mT : F = \cos v : \sin v \ ,$$

i.e.

$$mT = \frac{F}{\tan v} \ . \qquad (III.)$$

If we divide the equation (II.) by (III.) we obtain

$$\frac{mM}{mT} = \frac{FR^3 \cdot \tan v}{2F} \ ,$$

whence the independence of the deflection v, both of the magnetism of the needle m, and of the moment of rotation F, is evident of itself, and we have the following simple result:

$$\frac{M}{T} = \frac{R^3 \cdot \tan v}{2} \qquad (IV.)$$

"The determination of the intensity of the magnetism of the globe is therefore reduced to two principal operations.

"I. To observe the time of vibration of a needle NS, and to deduce from thence the moment of rotation which the terrestrial magnetism exerts on it."

This moment of rotation will be expressed by the product MT, and calculated by the equation (I.)

$$T = \frac{C}{M \cdot t^2} \ , \qquad \text{or} \qquad MT = \frac{C}{t^2} \ ,$$

in which C represents the moment of inertia of the bar, multiplied by the number, π^2, or 9.8696... and divided by the double of the space of the fall of a heavy body in the unit of time.

"II. A second needle, ns, being suspended: its position is observed; first, when subject to the influence of the earth's magnetism alone; and secondly, after NS has been placed at a considerable distance, as represented in the Figure. Then calculate from the difference between the two positions, or from the deflection, what fraction of the force of the earth's magnetism, the magnetic force of the needle NS, corresponds to at the selected distance. An equal fraction of the moment of rotation, found in (I.) gives the moment of rotation which the needle NS at that distance would impart to a similar one; this result, multiplied by the cube of the distance, gives the reduced moment of rotation; the square root of this gives the force of the needle NS in absolute measure: and finally, the number found in (I.) divided by this square root, gives the expression for the absolute measure of the earth's magnetism."

The ratio which the force of the bar on the needle (at the given distance R) bears to that of the earth's magnetism is expressed by the quotient

$$\frac{F}{mT} ,$$

and according to equation (III.)

$$mT = \frac{F}{\tan v} , \qquad \text{or} \qquad \frac{F}{mT} = \tan v \; .$$

But according to equation (II.)

$$mM = \frac{FR^3}{2} , \qquad \text{or} \qquad \frac{F}{mT} = \frac{2M}{R^3T} \; .$$

This fraction taken from the torque calculated according to equation (I.)

$$MT = \frac{C}{t^2} ,$$

i.e.

$$\frac{2M}{R^3T} \cdot MT = \frac{C}{t^2} \cdot \tan v ,$$

makes known to us the maximum moment of rotation which the bar with the magnetism M would exert on a similar bar at the distance R; for this maximum, according to the fundamental laws of magnetism, must be $\frac{2M^2}{R^3}$; and the above equation, gives

$$\frac{2M^2}{R^3} = \frac{C}{t^2} \cdot \tan v \; .$$

This result, multiplied by the cube of the distance R, gives double the reduced moment of rotation

$$2M^2 = \frac{CR^3 \tan v}{t^2} \; .$$

The square root of the half gives the force of the bar in absolute measure

$$M = \frac{1}{t}\sqrt{\frac{CR^3 \cdot \tan v}{2}} \; . \qquad (V.)$$

If, finally, we divide by the moment of rotation exerted by the earth on the needle, calculated according to equation (I.)

$$MT = \frac{C}{t^2} ,$$

we obtain

$$T = \frac{1}{t}\sqrt{\frac{2C}{R^3 \tan v}} , \qquad (VI.)$$

and this number expresses the absolute measure of the earth's magnetism.

"This appears the most easily understood exposition which can be given, without the use of mathematical signs, of the possibility of expressing the force of the earth's magnetism by a number which shall be perfectly independent of the individuality of the magnetic bars employed. In the actual application some points will appear in a somewhat different form, without, however, affecting in the least the nature of the method; and it will, besides, be necessary to take into consideration several collateral circumstances. We will add a few remarks on one or two circumstances.

"In speaking of the units to be employed in the measurements, mention was made only of a unit of distance and a unit of weight. But it should not be overlooked, that a certain weight, (a gramme, for instance,) does not mean, in this case, the quantity of ponderable matter which bears this name, and which is everywhere the same, — but the force which this quantity of matter exerts at the place of observation, under the influence of gravitation. It is well known that the force of gravity is not absolutely the same at different places; and if we chose the force of a gramme for our unit of weight, the intensity of the earth's magnetism would not be accurately measured by one standard at various places. The accuracy with which these measurements may now be made is such that this difference must not be neglected. The most simple way of meeting this difficulty is to reduce the force of gravity itself to an absolute quantity, by adopting as its measure the double height of descent in the unit of time, (for instance, a second,) and by expressing the force by the product of the mass into the number which measures the force of gravity. In this manner other numbers are obtained, both for the force of the magnetic needle employed, and for that of the earth's magnetism;[229] which numbers are based on three units, i.e., a unit of distance, a unit of time, and a unit of mass — instead of resting on the two units before spoken of."

In calculating the numbers M and T according to equations (V.) and (VI.)

$$M = \frac{1}{t}\sqrt{\frac{CR^3 \cdot \tan v}{2}} ,$$

$$T = \frac{1}{t}\sqrt{\frac{2C}{R^3 \tan v}} ,$$

the value ascribed to the constant C was

$$C = \frac{\pi^2}{g} \cdot K ,$$

in which π represented the known number 3.14159...; g, double the space of descent in the unit of time; K, the moment of inertia of the vibrating bar. The new numbers are obtained from the same equations, by ascribing to C the value

$$C = \pi^2 K .$$

"One main difficulty in the application of this method consists in the fact, that the above-mentioned law holds good, namely, that the action of a magnetic needle is

[229][Note by CFG:] They are to the previous numbers in the proportion of the square root of the number which measures the force of gravity to unity.

inversely as the cube of the distance, with sufficient accuracy only for very great distances, and where the effects are too small to be determined with precision by direct observation. At moderate distances the variations from the law become very perceptible; but theory teaches that these very differences are subject to rule; and mathematics afford us the means of recognising, and almost wholly eliminating them, by the combination of experiments made at *various* moderate distances."

For the purpose of showing the application of the small measuring apparatus to the above-mentioned observations, we shall give lastly, in a few words, the necessary process of correction. This is threefold:

1. Instead of the values given by direct observation for the deflexions v_0, v_1, v_2, etc., of the needle by the magnet bar acting at various distances, R_0, R_1, R_2, etc., the following combined values are to be taken:

$$\begin{aligned} v_0 &= \frac{1}{4}(u_0 - u_0' + u_0'' - u_0''') \\ v_1 &= \frac{1}{4}(u_1 - u_1' + u_1'' - u_1''') \\ v_2 &= \frac{1}{4}(u_2 - u_2' + u_2'' - u_2''') \\ &\textit{etc.} \end{aligned}$$

2. To the approximate values of M/T, which were obtained by equation (IV.)

$$\frac{M}{T} = \frac{R^3 \tan v}{2},$$

the following corrections are added:

Approximate value of M/T.	Correction.
$\frac{R_0^3 \tan v_0}{2}$	$-\frac{L}{R_0^2}$
$\frac{R_1^3 \tan v_1}{2}$	$-\frac{L}{R_1^2}$
$\frac{R_2^3 \tan v_2}{2}$ etc.	$-\frac{L}{R_2^2}$ etc.

3. As the number of the measured dimensions R_0, R_1, R_2, etc. and v_0, v_1, v_2, etc., is greater than is required for the determination of the unknown quantities L and M/T, the rules of the calculus of probabilities are employed in order to obtain from them the most probable values of those quantities. These rules are as follows.

From the quantities R_0, R_1, R_2, etc., v_0, v_1, v_2, etc., we must calculate the following expressions:

$$\frac{\tan v_0}{R_0^3} + \frac{\tan v_1}{R_1^3} + \frac{\tan v_2}{R_2^3} + \text{etc.} = A\,,$$

$$\frac{\tan v_0}{R_0^5} + \frac{\tan v_1}{R_1^5} + \frac{\tan v_2}{R_2^5} + \text{etc.} = A'\,,$$

$$\frac{1}{R_0^6} + \frac{1}{R_1^6} + \frac{1}{R_2^6} + \text{etc.} = B\,,$$

$$\frac{1}{R_0^8} + \frac{1}{R_1^8} + \frac{1}{R_2^8} + \text{etc.} = B' ,$$

$$\frac{1}{R_0^{10}} + \frac{1}{R_1^{10}} + \frac{1}{R_2^{10}} + \text{etc.} = B'' ;$$

thence we shall have the most probable value of[230]

$$L = \frac{1}{2} \cdot \frac{AB' - A'B}{B'^2 - BB''} ,$$

$$\frac{M}{T} = \frac{1}{2} \cdot \frac{A'B' - AB''}{B'^2 - BB''} = r .$$

From this, and equation (I.)

$$MT = \frac{C}{t^2} ,$$

we obtain

$$M = \frac{1}{t}\sqrt{rC} , \qquad (VII.)$$

$$T = \frac{1}{t}\sqrt{\frac{C}{r}} . \qquad (VIII.)$$

The experiments with the small measuring apparatus may be calculated according to these laws and formulae, and the absolute magnetism of the bar and that of the earth determined.

10.4 Calculation, According to the Above Rules, of the Observations Made with the Small Measuring Apparatus

The experiments were, 1st, those of deflection, which gave the values of $u_0 - u_0'$, $u_1 - u_1'$, $u_2 - u_2'$, $u_2'' - u_2'''$, $u_1'' - u_1'''$, $u_0'' - u_0'''$, and the corresponding values of R, viz. R_0, R_1, R_2, R_2, R_1, R_0. We calculate from these the values of v_0, v_1, v_2, corresponding to R_0, R_1, R_2; and hence the values of A, A', B, B', B'', which are simple functions of the six quantities, v_0, v_1, v_2; R_0, R_1, R_2. And lastly, the value of r is deduced from the quantities A, A', B, B', B'', of which it is a function. Thus, the value of r is obtained by calculation from the experiments of deflection.

230[Note by AKTA:] In the English translation the next two equations were written as:

$$L = \frac{1}{2} \cdot \frac{AB' - A'B}{B'^2 - B''^2} ,$$

$$\frac{M}{T} = \frac{1}{2} \cdot \frac{A'B' - AB''}{B'^2 - B''^2} = r .$$

2nd. From the experiments of vibration the value of the time of vibration t is found: having thus the values of r and t, it is only required, for the purposes of the travelling observer, to calculate

$$\frac{1}{t\sqrt{r}}\ ;$$

for this value is proportional to the number, which expresses the absolute terrestrial magnetism, and consequently suffices for the *comparison* of the absolute intensity of all places where such experiments may be performed. Such a *comparison* is usually the only object sought by the travelling observer. It may sometimes, however, be desirable to obtain not only comparisons of the absolute intensity at various places, but the absolute intensity itself; the apparatus may be lost on a voyage, and be replaced by a new one; and it then becomes necessary, in order to compare the two series of results obtained with instruments which cannot be compared together, to calculate the moment of inertia of the magnet bar, the time of vibration of which had been observed, and to extract its square root. The product of the quantity $\frac{1}{t\sqrt{r}}$ into the square root, and into the number $\pi = 3.14159...$ gives a number expressing the earth's magnetism in absolute measure.

On this account it is advantageous that the bar should be an accurate parallelopiped, because in such case the moment of inertia can be deduced for the present purpose directly from the weight p, the length a, and the breadth b of the bar. For it is well known that the square $a^2 + b^2$ of the diagonal of the superficies of the parallelopiped, multiplied by the mass p of the weight, and divided by 12, gives the moment of inertia sought, in the case in which the bar shall have been suspended by the centre of that superficies. Consequently in the equations (VII.) and (VIII.)

$$C = 9.8696 \cdot \frac{a^2 + b^2}{12} \cdot p\ .$$

If we compare the observations above mentioned with these formulae, it will be seen that the following quantities have been directly measured, and the following values found for them:

$$\begin{aligned} u_0 - u_0' &= 23°9' \\ u_1 - u_1' &= 47°42' \\ u_2 - u_2' &= 71°48' \\ u_2'' - u_2''' &= 69°21' \\ u_1'' - u_1''' &= 46°12' \\ u_0'' - u_0''' &= 22°27' \end{aligned}$$

$$\begin{aligned} R_0 &= 450\ mm \\ R_1 &= 350\ mm \\ R_2 &= 300\ mm \end{aligned}$$

$$t = 6.67''$$

$$a = 101.0\ mm$$
$$b = 17.5\ mm$$

$$p = 142000 \text{ milligrammes.}$$

From these may next be calculated,

$$v_0 = \frac{1}{4}(23°9' + 22°27') = 11°24.00'$$

$$v_1 = \frac{1}{4}(47°42' + 46°12') = 23°28.50'$$

$$v_2 = \frac{1}{4}(71°48' + 69°21') = 35°17.25' \ .$$

If now we take the second and the millimetres as the fundamental units of time and space in our calculation, we may deduce from the ascertained values of R_0, R_1, R_2, v_0, v_1, v_2, the following values of A, A', B, B', B'', viz.

$$A = \frac{\tan 11°24'}{450^3} + \frac{\tan 23°28.5'}{350^3} + \frac{\tan 35°17.25'}{300^3} = \frac{385.54}{10^{10}} \ ;$$

$$A' = \frac{\tan 11°24'}{450^5} + \frac{\tan 23°28.5'}{350^5} + \frac{\tan 35°17.25'}{300^5} = \frac{384.86}{10^{15}} \ ;$$

$$B = \frac{1}{450^6} + \frac{1}{350^6} + \frac{1}{300^6} = \frac{2.0362}{10^{15}} \ ;$$

$$B' = \frac{1}{450^8} + \frac{1}{350^8} + \frac{1}{300^8} = \frac{2.0277}{10^{20}} \ ;$$

$$B'' = \frac{1}{450^{10}} + \frac{1}{350^{10}} + \frac{1}{300^{10}} = \frac{2.0855}{10^{25}} \ .$$

From these r is calculated:[231]

$$r = \frac{1}{2} \cdot \frac{385.54 \cdot 2.0855 - 384.86 \cdot 2.0277}{2.0362 \cdot 2.0855 - (2.0277)^2} \cdot 10^5$$

or[232]

[231] [Note by AKTA:] In the English translation the next equation appeared as:

$$r = \frac{1}{2} \cdot \frac{385.54 + 2.0855 - 384.86 + 2.0277}{2.0362 + 2.0855 - (2.0277)^2} \cdot 10^5 \ .$$

[232] [Note by AKTA:] In the original German text we have $r = 8765000$. Due to a misprint in the English translation, [Web41a], this value appeared as $r = 87650000$. We inserted here the original value.

$$r = 8765000 \ .$$

Finally, from this value of r, and from that of t, determined by observation, may be deduced the value:

$$\frac{1}{t\sqrt{r}} = \frac{1}{6.67 \cdot \sqrt{8765000}} = \frac{5.0641}{10^5} \ .$$

This number suffices for the comparison of all intensities measured with the same instrument, however the magnetic condition of the apparatus may have varied.

Further, the number T, which expresses in absolute measure the resulting intensity of the earth's magnetism, may be ascertained by deducing from the observations the value of C, and multiplying the former number by its square root. C is calculated from the observed values of a, b, and p, the mass of the milligramme being taken as the unity of mass:[233]

$$C = 9.8696 \cdot \frac{101^2 + 17.5^2}{12} \cdot 142000 = 0.12272 \cdot 10^{10}$$

whence T is deduced

$$T = 5.0641 \cdot \sqrt{0.12272} = 1.774 \ .$$

10.5 Examination of the Result

This number 1.774, expressing the intensity of terrestrial magnetism on the 18th of January, 1837, possesses, as an absolute measure, the advantage of being directly comparable with the results obtained in July 1834 with the magnetometer of the Göttingen Magnetic Observatory, published in the *Göttingen gelehrten Anzeigen* of that year. They will be found in part 128,[234,235] (with the account of the newly-constructed building, and of the instruments, as well as of the first experiments performed there). They are as follow:

July 17	1.7743
July 20	1.7740
July 21	1.7761.

Two apparatus destined for the same purpose can hardly be more dissimilar than the small apparatus above described, and the magnetometer. It results from the comparison, that the intensity of the terrestrial magnetism in Göttingen has undergone hardly any alteration from 1834 to 1837.

We have also a direct comparison of this number obtained for Göttingen with the result of observations with a third apparatus, differing widely from both the others made at Munich,

[233][Note by AKTA:] Due to a misprint in the English translation, [Web41a], the next equation appeared as

$$C = 9.8696 + \frac{101^2 + 17.5^2}{12} + 142000 = 0.12272 + 10^{10}$$

.

We replaced it by the original equation.

[234][Note by ER:] Gauss' *Werke*, Vol. V, p. 524.

[235][Note by AKTA:] [Gau34a, p. 524 of Gauss' *Werke*].

April 1st, 1836, viz. 1.905, and with the number found for Milan, with the magnetometer of that place, in October, 1836, viz. 2.01839.

To gain a clear idea of the import of these numbers, the determination and application of which have been hitherto under consideration, imagine a number of small steel bars, perfectly alike, and each weighing about $2\frac{1}{2}$ grammes, or $\frac{1}{6}$ of an ounce. Imagine further a balance, of which the length of the arms bears to 1 metre the same proportion that 1 metre bears to the space of descent in 1 second (204 millimetres nearly); suppose one of these steel bars to be attached in a parallel direction to the horizontal beam of the balance, in such manner that the equilibrium is not thereby disturbed. Then render all the steel bars (including the one attached to the balance) *equally magnetic*,[236] and to such a degree that when another of their number is placed vertically beneath the scale at the distance of 1 metre from the attached magnet bar, $\frac{1}{1000}$th of a milligramme must be placed in the scale to preserve equilibrium. When the magnetism of all the bars has been regulated in this manner, place one of the bars horizontally, and at right angles to a small compass needle, 1 metre from the centre of the needle beneath, taking care that as the compass needle is deflected from the magnetic meridian, the bar be also turned so that they may preserve their rectangular position. Lastly, calculate how many such bars are required that their united force may deflect the compass needle 90°; the number of bars gives the terrestrial magnetism in thousandths of its absolute measure.

We may conceive in like manner the number which represents the absolute measure of the terrestrial magnetism to represent the number of these bars reckoned in thousands, the forces of which must be united to cause, at a distance of a metre, a deviation of 90°. This would require at

Göttingen the force of 1775 bars,
Munich the force of 1905 bars,
Milan the force of 2018 bars.

10.6 On the Advantages of the Dimensions Selected for the Small Measuring Apparatus

Before concluding this article, we have to discuss the accuracy of which the absolute measurement of intensity with the apparatus described is susceptible, and on what it is founded. It has been already remarked, that the absolute intensity can be measured with the accuracy it deserves only with the magnetometer. It is therefore unnecessary to state that such extreme accuracy cannot be attained with the small apparatus. And in order to obtain with it a good approximation, it must combine all the advantages of which it is susceptible.

The difficulty of an accurate measurement of intensity, with other instruments than the magnetometer, is thus stated in the memoir "*On Terrestrial Magnetism and the Magnetometer*:"[237,238]

> "In all cases, if the elimination is to be satisfactory, the experiments must not be performed at too small distances; consequently the effects are always comparatively

[236][Note by AKTA:] In German: *gleich stark magnetisch*. That is, all bars magnetized with the same strength.

[237][Note by ER:] Gauss' *Werke*, Vol. V, p. 332.

[238][Note by AKTA:] [Gau36b, p. 332 of Gauss' *Werke*].

small, and the means previously in use are inadequate to measure them with the necessary precision. It is this difficulty which has called for, and has given rise to the construction of a new apparatus, which may with propriety receive the name of *magnetometer*, since it serves to execute, with an accuracy equaling that of the most delicate astronomical determinations, all measurements — both of the force of magnetic needles, and of the intensity of the earth's magnetism (at least its horizontal portion). — The (horizontal) direction of the earth's magnetic force is determined accurately with it to within one or two seconds of arc; the commencement and termination of a vibration is observed with it to within a few hundredths of a second of time, and consequently more accurately than the passage of stars behind the wires of a transit."

There are two circumstances, chiefly, on which the accuracy of an absolute measurement of intensity depends; *first*, the magnitude of the deflection produced; *secondly*, the delicacy of the instrument in measuring this deflection. In constructing an apparatus for this purpose we may therefore follow two different paths: we may either make the amount of deflection the main object, and pay only as much attention to the means of measurement as may be consistent therewith; — or we may attend chiefly to accuracy in the means of measurement, and let the amount of the deflection be the second object. The latter plan leads to much greater accuracy than the former, for this reason: the amount of deflection soon attains a limit, on account of the necessary condition of a considerable distance between the deflecting bar and the needle, so that the deflection produced must always be small. If, however, all pretensions to great accuracy of measurement are relinquished at the outset, by making the magnetic needle play on a pivot, instead of suspending it by a silk thread, the friction of the point renders fineness of measurement quite illusory, and the former much less advantageous plan is the only one that remains open; the endeavour must then be to adopt the arrangements and proportions best suited to produce the greatest possible deflection.

This is the express object of the small size of the apparatus described, and not merely to render it light and convenient of transport.

That the small size of the apparatus does actually allow of a great amount of deflection is evident by the result; for in the experiments above mentioned all the measured angles exceeded 22°: it is easy to explain the reason.

1. The distance of the deflecting bar from the needle must be *relatively* great, but need not be *absolutely* so: it must at least be three or four times greater than the length of the deflecting bar, or of the magnetic needle.

2. By diminishing in proportion *all* the linear dimensions of the apparatus (viz. the dimensions of the magnets, and their distance apart), the angular magnitudes, of which the deflection is one, remain unchanged; therefore such proportional reduction in the size of the apparatus causes no loss in the amount of the deflection to be measured.

3. But if instead of diminishing in equal proportion *all* the linear dimensions of the apparatus, we diminish only the length of the magnets and their distance apart, the breadth and thickness of the deflecting bar being little or not at all diminished, then we even gain an increase in the angular magnitudes, and it only remains to know how far this increase may be carried.

The limit depends on a single circumstance, viz. on the breadth and thickness of the deflecting bar, with a given length. Experience has shown, that neither the breadth nor the thickness of the bar ought to exceed the eighth part of its length. It follows that the greatest deflection may be produced by a magnet bar, of which the breadth and the thickness are equal, and of which the length is eight times greater than either, and acting upon a magnetic needle, placed at a distance equal to three or four times the length of the bar; the length of the needle must not exceed that of the bar.

From this rule then we obtain the most advantageous dimensions of such an apparatus, by knowing the limit in respect to thickness, which is determined by the *nature of the steel*.

The thickness of the bar must not amount to much more than $12\frac{1}{2}$ millimetres, as otherwise the steel cannot be properly hardened and magnetized throughout. We thence obtain the following dimensions of the deflecting bar, as those which combine the greatest advantages, namely, for its breadth and thickness $12\frac{1}{2}$ millimetres, and for its length 100 millimetres. We have also the length of the magnetic needle 100 millimetres, and the smallest admissible distance between them, 300 millimetres.

By following these rules we obtain an apparatus, with which, in mean latitudes, the smallest deflections to be measured exceed 22°, as in the experiments related. At greater distances from the magnetic poles of the earth, this deflection becomes somewhat smaller; nearer to the magnetic poles it is much larger. Therefore, if these deflections can be accurately measured to within a tenth part of a degree, a final result can be obtained to within the 200th part of the force itself; since all other measurements required in the determination of the absolute intensity can be made with greater accuracy. This result, it is true, is far inferior to that which can be obtained with the magnetometer; but such results may still be of great utility in the absence of more accurate determinations.

Chapter 11

[Gauss, 1838] On a New Instrument for the Direct Observation of the Changes in the Intensity of the Horizontal Portion of the Terrestrial Magnetic Force

Carl Friedrich Gauss[239,240,241]

It is well known that for the perfect determination of the terrestrial magnetic force at a given place, *three* elements are required; and, in general, the Declination, Inclination, and Intensity are selected for the purpose. Although this choice is the most simple in conception, it is not only allowable, but in many respects it may be advisable, to adopt a different combination. In practical as well as in theoretical respects, it is far more advantageous to consider the horizontal portion of the terrestrial force separately, and to imagine it in two elements, the direction (declination) and the intensity. If we add to these, as a third element, — either the intensity of the vertical force, or the inclination, — the intensity of the total force, if desired, may be directly obtained.

With respect to the two elements of the horizontal force, with which alone we are here concerned, all the questions which occur in regard to the *declination* are completely met by the magnetometer which has been in use since 1833.[242,243] This instrument serves with a certainty, convenience, and accuracy that leaves nothing more to be wished; not only for the determination of the absolute value of the declination, but also for following its regular and accidental changes, from year to year, from month to month, from hour to hour, — nay,

[239] [Gau38b] with English translation in [Gau41c].

[240] The Notes by C. F. Gauss are represented by [Note by CFG:]; the Notes by Richard Taylor, the editor of the *Scientific Memoirs* where this translation has been published, are represented by [Note by RT:]; while the Notes by A. K. T. Assis are represented by [Note by AKTA:].

[241] [Note by CFG:] This essay contains the essential content of the lecture I gave at the public meeting of the Royal Society of Sciences (*Königlichen Societät der Wissenschaften*) on September 19, 1837.

[242] [Note by RT:] *Scientific Memoirs*, part V, pp. 25 *et seq.*

[243] [Note by AKTA:] [Web37a] with English translations in [Web41f], [GW66] and [GW39a]; and French translation in [Web38c].

from one minute to the other. This magnetometer also determines, in absolute measure, the *intensity* of the horizontal portion of the earth's magnetic force, — which was, in fact, the object which first gave rise to its construction: it does not, however, by any means, solve this problem *perfectly* in *all* respects.

The application of the magnetometer to determine the magnetic intensity is founded on a combination of *several* operations, one of which consists in observing the time of vibration[244] of a needle. But this operation, from its very nature, requires a considerable time, as the number of vibrations from which we deduce the duration of a single vibration ought not to be too small. Now, supposing the magnetic intensity to be constant during the period employed in the observation, the resulting time of vibration will correspond *truly to the intensity*; but if the latter has varied in the interval, the time of vibration will only correspond to *its mean value*. Whatever changes may have taken place *during* the interval are entirely concealed from us, the instrument giving only average values. If, in order to approximate more closely, we were to choose shorter intervals, or to base the results upon a smaller number of vibrations, we should sacrifice accuracy and certainty, and be in danger of considering errors of observation as anomalies in the intensity.

But the more interesting the magnetic disturbances in short intervals appeared, — as shown by the experiments of last year, in regard to the declination only, — the more important it was to possess a means by which the effects of similar disturbances in the intensity might be followed and measured with the same ease, certainty, and accuracy.

We have already seen that the unfitness of the method hitherto employed for *this* purpose consists in the circumstance, that it is based on observations of the times of vibration, which, from their very nature, must always require a long interval. Now the time of vibration serves in this case only to determine indirectly the moment of rotation[245] which the earth's magnetism imparts to the needle when it is not situated in the magnetic meridian. If, then, we can determine accurately this moment of rotation in a direct manner without observations of vibration, and if we can measure its variation with accuracy, quickness, and certainty, our main object is attained. The method to be described for this purpose rests on the following basis.

The necessary conditions of equilibrium of a body of any form suspended to *two threads*, — its parts being supposed, in the first instance, subject to gravity alone, and firmly connected with each other, — may be thus briefly described: the vertical passing through the centre of gravity of the body, and the straight lines coinciding with the threads, are in one plane, and are either parallel with each other, or intersect in a fixed point. In all cases, therefore, in the position of equilibrium, the two threads and the centre of gravity are in one vertical plane. To give precision to our ideas, it may be supposed that the two threads are of equal length; that the two upper points of connexion are at the same height, and that their distance apart is the same as that of the two lower points; and lastly, that the two latter form with the centre of gravity an equiangular triangle. Under these suppositions, in a state of equilibrium, the two threads will hang vertically, and a third vertical line, midway between them, will pass through the centre of gravity of the body. If we remove the body from this position by means of a rotation around the last-named line, the two threads will no longer be vertical, nor will they be in one plane, and at the same time the body will be somewhat raised. There arises consequently a tendency to return to the former position, with a moment of rotation, which may, with sufficient accuracy, be regarded as proportional

[244][Note by AKTA:] In German: *Schwingungsdauer*.
[245][Note by AKTA:] In German: *Drehungsmoment*.

to the sine of the deviation from the position of rest, and which is, therefore, greatest when the deviation amounts to 90 degrees. This maximum effect is always understood when the moment of rotation is spoken of; it may also be regarded as the force by which the body is retained in equilibrium by its mode of suspension, and which, for shortness, I shall call the directive force of suspension.[246] The magnitude of this force depends, 1st, on the length of the suspending threads; 2nd, on their distance apart; 3rd, on the weight of the body; being inversely proportional to the length of the threads, and directly as the square of their distance apart, and as the weight of the body. If the above suppositions are not fulfilled, the expression for the directive force is more complicated, and the reaction of the threads against the torsion also renders a small modification necessary. Means are not wanting to enable us to determine by experiment, with the greatest accuracy, the magnitude of the directive force. If the body is left to itself, after having been made to deviate from its position of equilibrium, it will vibrate with the greatest regularity, the middle of the vibrations coinciding with this position, and the duration depending on the magnitude of the directive force and on the moment of inertia of the body.

If we further suppose a horizontal magnet bar to form a part of the suspended body, a second directive force is exerted, and the phaenomena depend on the combinations of the two forces, according to the known laws of statics. There are, in this point of view, three cases to be distinguished, according as the two positions of the body, in which it would be in a state of equilibrium arising from either of the two forces acting singly, either coincide, — or are opposite, — or form an angle with each other. It is easily seen that the difference between these three cases rests on the relation of the two angles, which the straight line joining the two lower points of connexion of the thread forms with the magnet bar; and which the straight line joining the two upper points of suspension forms with the magnetic meridian. If we imagine the body in that position of equilibrium which is due solely to the mode of suspension, the magnet bar must be, in the first of our three cases, in the magnetic meridian, and in its natural position (*i.e.* the north pole towards the north); in the second case, it must be in the magnetic meridian, but in the reverse position; and in the third case, it must form an angle with the magnetic meridian. For the sake of brevity, I will call these three positions of the magnet bar, the direct, the reverse, and the transverse positions.

In the direct position, the action of terrestrial magnetism on the magnet bar does not change the position of equilibrium corresponding to the mode of suspension; but the apparatus is retained in the same position by an increased force, which is the sum of the two directive forces.

In the reverse position, the equilibrium does not cease, but it is only stable when the magnetic directive force is smaller than the directive force arising from the mode of suspension; and the body is then only retained in this position by a force which is the difference between the two directive forces. If the magnetic directive force were the greater, the equilibrium would be unstable, and the body once disturbed from that position would not return to it, but would depart further and further from it, and only come to rest in the opposite position, in which the bar is in its natural position in space, but the suspending threads cross one another.

Finally, in the third case, where the two directive forces form an angle with each other, the conflict of these forces will end in an intermediate position, where, on the one hand,

[246][Note by AKTA:] In German: *Man kann dasselbe auch als das Maass einer Kraft ansehen, mit welcher der Körper vermöge der Aufhängungsart in seiner Gleichgewichtsstellung zurückgehalten wird, und die ich der Kürze wegen die aus der Aufhängungsart entspringende Directionskraft nennen will.*

the bar will not be in the meridian, and, on the other, a straight line through the lower points of connexion of the threads will not be parallel to a straight line through the upper points. This intermediate position, and the force by which the apparatus is retained in it, obey the statical laws of the composition of two forces. It will now be easily seen, that if the apparatus presents the means of measuring the angle between the three positions in question, the relation of the two component directive forces may be calculated, and consequently, we can express in absolute measure the magnetic directive force, if the force arising from the mode of suspension is also known in absolute measure. Thus our problem is solved. It is most advantageous to dispose the magnet bar relatively to the other parts of the apparatus, so that it shall form, in the mean position of equilibrium, nearly a right angle with the magnetic meridian, to which case the term *transverse* position is chiefly applicable. By this means, the deviation of the threads from their position in one plane is greatest, and the result is therefore most accurate; and a small change in the magnetic declination, arising from hourly or accidental fluctuations, has no perceptible influence on the position. On the contrary, every change in the intensity of terrestrial magnetism affects the position directly, and can at once be recognised and measured with the same ease, quickness, and accuracy as the changes of declination are by the ordinary magnetometer.

I had many years ago ascertained the practical applicability of this mode of determination, by preliminary experiments with an apparatus (very rough it is true) to which I have alluded in my Memoir on Terrestrial Magnetism and on the Magnetometer.[247] Recently, however, I have had a more perfect apparatus constructed, and have suspended it in the astronomical observatory, in the spot previously occupied by the magnetometer with the bar of 25lbs. weight. After what has been said, a few words will suffice for the description of this apparatus. It is suspended by two steel wires 17 feet in length, or, to speak more accurately, by a single one, the extremities of which are attached to the apparatus beneath, whilst, above, the centre of the wire passes over two cylinders which keep it at a proper distance apart (about $1\frac{1}{2}$ inch); by this arrangement the two wires have, of themselves, an equal tension. The suspension is above the ceiling of the room, and the wires hang freely through a circular aperture ($3\frac{1}{2}$ inches wide) in the ceiling. The interval of the wires, both above and below, can be increased or diminished at pleasure. The apparatus suspended to the wires consists of four principal parts. The first, to which the wires are affixed, is an horizontal circular disk, 4 inches in diameter, divided on silver into quarter degrees. The second part consists of an alidade, concentric with the circle, and rotating on its limb, and having two verniers indicating minutes; a strong rod, perpendicular to the plane of the circle, is firmly connected with the alidade, and to this is fixed a very perfect circular mirror $1\frac{1}{2}$ inch in diameter, in which may be seen, through a telescope placed at the distance of 16 feet, the image of a portion of an horizontal scale, divided in millimetres, fixed below the telescope. In this manner every change in the position of the circle may be seen and measured; small changes directly, and with great accuracy by the divisions of the scale seen in the telescope; and greater ones by combining a movement of the alidade and reading off the verniers. The third portion of the apparatus is a stirrup situated beneath the circle, being a double frame, in which the fourth constituent part, a 25-pound magnet bar, is inserted. This stirrup has likewise a rotatory motion round the centre of the circle, and is provided with two indexes applied to the limb of the circle, by which the amount of the rotation can be measured to a minute.

If now we place the stirrup so that the apparatus preserves the same position of equilib-

[247][Note by AKTA:] [Gau36b, p. 19 of Schumacher's *Jahrbuch für 1836* and p. 327 of Gauss' *Werke*].

rium, whether the magnet bar be in the stirrup, or a non-magnetic body of equal weight, we have the first or the second of the positions distinguished above, according as the magnet bar is in the direct or in the reverse position. The first affords no particularly important practical application; and the advantage of the second is connected with the condition, that the magnetic directive force be somewhat less than the directive force due to the mode of suspension. With our apparatus the proportion of these forces is at present nearly as ten to eleven: the resulting directive force is consequently only the tenth part of the magnetic directive force. We have consequently in this case an arrangement analogous to an astatic magnetic needle;[248] and every extraneous force that disturbs the direction of the ordinary needle is indicated here by a tenfold greater effect than would take place in the case of suspension by a single thread; and, as will be easily perceived, in the opposite direction. This then affords, among other things, the solution of a problem which has been often attempted without success, viz. that of representing the daily and hourly changes of the magnetic declination under a magnified form. Numerous simultaneous observations of this kind, made with this apparatus, and with the magnetometer of the observatory, have afforded the most satisfactory results. This application, however, has lost much of its importance from the introduction of the declination magnetometer, which gives the minutest changes with all the accuracy that can be desired.

This and other applications of the instrument, with the bar in the reverse position, to which I shall hereafter return, must, however, be considered as of minor importance; the employment of the apparatus in the third or transverse position for observations of the intensity being far more important. If in proceeding from the direct position the magnetic bar is deflected from the magnetic meridian by turning the stirrup, the whole apparatus, in order to regain its equilibrium, must turn back through a certain angle corresponding to the proportion of the two directive forces; the difference of the two angles will be the deviation of the magnetic bar from the magnetic meridian when in the position of equilibrium; and it may easily be arranged so that this deviation shall amount to nearly 90 degrees, and thus the advantages previously spoken of be gained. In this position the apparatus is peculiarly well adapted for observing changes of the intensity, which are immediately indicated by changes of position. In regard to such changes as only take place in long intervals, several circumstances must be attended to; for instance, it is requisite that from time to time we examine, by known and appropriate means, whether and to what extent the magnetism of the bar may have changed; the variations of temperature must also be considered, both in their effect on the magnetic state of the bar, and on the interval and length of the suspending wires, and, consequently, on the directive force arising from the mode of suspension. But with respect to the irregular changes of the intensity in short intervals, this apparatus performs the same service as the magnetometer does in respect to similar changes of the declination; and the mode of observation with both instruments is the same. The changes of intensity are obtained, expressed in parts of the scale, which, however, may easily be reduced to fractions of the intensity itself. Under the present relations of the apparatus, the 22,000th part of the entire intensity answers to one division of the scale.

The experiments hitherto made with the apparatus, comprising as yet but a very short period, have already indicated some important results.

In the first place, the observations indicate the regular changes dependent on the time of day, which, it is true, are as frequently intermixed with irregular ones as in the declination; to discriminate between them with certainty will require observations continued for years. If I

[248][Note by AKTA:] See footnote 150 on page 43.

may venture, from the very little experience hitherto gained, to express a supposition rather than a result, the regular change seems to consist in this, — that the intensity decreases in the hours of the forenoon, so that it attains its minimum one or two hours before mid-day, and then again increases. In order, however, to obtain provisionally the quantitative ratio, I have noted the position in the morning at 10, and in the afternoon at 3, on thirty days in August, 1837. The result was, that on twenty-six days, the intensity was greater in the afternoon than in the morning, and less on only four days; the mean difference amounting to 39 parts of the scale, or somewhat more than the 600th part of the entire intensity. On most of these days the apparatus was also noted in the morning at 9 o'clock; of twenty-eight days, there were twenty-three on which the intensity was still at this hour greater than an hour later, and the reverse was found to be the case on five days only; the mean difference, however, amounted in this case only to $11\frac{1}{2}$ divisions of the scale, or somewhat more than the 2000th part of the entire intensity.

Secondly, several very extensive series of observations prove that irregular, and, at times, very considerable disturbances, and varying in short intervals of time, occur not less frequently in the intensity than in the declination, as analogy would have led us to expect. Uninterrupted series of some length have been made on these occasions simultaneously with the intensity apparatus, and with the magnetometer of the observatory; 1st, on the 15th July, 1837, from 6 A.M. to 6 P.M.; 2ndly, in the usual magnetic term of the 29th and 30th July; and 3rdly, during the extra term of the 31st August and 1st September; the observations being made at every five minutes. In comparing the two series, it is observable, that where the declination was violently disturbed, in general great disturbances also occur in the intensity.[249]

By the representation of the changes of the declination and of the intensity in two distinct curves, as is done for the November term in the first Part of the Plate,[250] we are far from obtaining so perfect an image of the course of the disturbances as by their combination in a single curve.

---

[249] [Note by CFG:] In a similar way, and with the same result, observations were also subsequently made with both apparatus in the term of 13th to 14th November.

[250] [Note by RT included at the end of this English translation:] The graphical representations of the changes in the direction and intensity of the horizontal force at Göttingen in the terms of July 29-30, Aug. 31—Sept. 1, and Nov. 13-14, 1837, referred to in the preceding memoir, page 259 [of Volume 2 of Taylor's *Scientific Memoirs* where this English translation was published], are contained in the *Resultate aus den Beob.* for 1837. As our purpose is rather that of illustration than of record, it has appeared sufficient to give one of these plates; and we have selected for the purpose that of the term of November 13-14, appointed expressly on those days, on account of the great number of *falling stars* which had been observed in them in preceding years. Part 1 of the Plate, represents the changes of the Intensity in the upper line, and of the Declination in the lower. The justice of the remark, in p. 259, will be at once recognised, namely, that, when considerable changes take place in the one element, they are usually accompanied by considerable changes in the other. Part 2 represents the changes of both elements united in one curve, and affords an illustrative delineation of the variation of the horizontal portion of the earth's magnetic force. To avoid the confusion arising from too repeated involutions of the curve, it is divided into three separate portions, and in each of these half the curve is drawn in an unbroken line, and half in a dotted line.

M. Gauss remarks, that "the observations during this term of Nov. 13-14 do not present greater disturbances than had been noticed in many of the terms at other seasons of the year. On the preceding and following evenings, very great and rapidly varying changes took place in the declination; but these are known to be the general accompaniments of the Aurora Borealis, which was extremely brilliant on those two nights." — Edit.

A complete representation of the terrestrial magnetic force (*i.e.*, its horizontal portion) at each moment, is given by a single straight line, of which the length is proportional to the intensity, and the angle which it forms with a fixed line is equal to the declination. To represent the force in several successive moments, the same point of commencement of the different straight lines is preserved, and the terminal points alone exhibited; these are noted with the corresponding times, and are united by a line (Plate, part 2). The radii themselves are not drawn, and even the common point of commencement, to obtain anything like a convenient scale, must always be situated far beyond the drawing. This leads us to a new point of view, from which we may consider such changes of the two magnetic elements. In fact, they are the two horizontal components of that always comparatively small disturbing force, to which the mean terrestrial magnetic force is at each moment subject, resolved into two directions — one in the magnetic meridian, and the other perpendicular to it. The second component is given directly by the magnetometer, the first by the new apparatus; for which reason both must be reduced to a common measure before the drawing.

In applying this very illustrative mode of representation, it must be remembered that the course of the changes during a whole day cannot be represented in one drawing without confusion, if there are frequent and quickly varying changes, as the curve would present too many convolutions: it is necessary, therefore, in such case, to make separate drawings for shorter intervals.

If we compare the new apparatus and the magnetometer, we find that the two, with respect to *some purposes*, serve reciprocally to render each other complete; but, in other respects, have one and the same application. For the determination of the absolute declination, the magnetometer alone is applicable, and not the new apparatus. The changes of the declination, and especially the quickly varying changes, may be followed with both. For determining the absolute intensity, both apparatus may be employed, although the use of the magnetometer is somewhat less complicated than the sole employment of the new instrument would be; but the former of itself can only give the mean value of the intensity for a certain interval, and the quickly varying changes are entirely overlooked with this instrument, while the new instrument indicates them most satisfactorily. For all other purposes — for instance, for comparing magnetic bars with one another in respect to their magnetic strength, — and in connexion with a multiplier, for galvanometrical and telegraphical purposes, — both are alike useful. With respect to the two last applications, the new apparatus has an important advantage, in its being in our power, as above mentioned, to render it as nearly astatic as may be desired. A few instances of the sensibility of the apparatus as a galvanometer may be here noticed. The multiplier surrounding the magnetic bar contains 610 coils of copper wire covered with silk, and the galvanic current has to pass through a length of wire of more than 6000 feet. This length increases to 13,000 feet if the current at the same time is brought from the physical cabinet. In general, however, other apparatus are brought into connexion with the chain,[251] so that in many experiments the whole length of wire amounts to 40,000 feet, or nearly two German miles. By far the greatest portion of this wire is very thin; and this length, in so far as the force of the current is affected by it, is equivalent to a wire about eight German miles in length, of the thickness of the connecting wire between the Astronomical Observatory and the Physical cabinet. Notwithstanding this long chain, even the weakest galvanic forces give the heavy magnetic bar a deflection not merely perceptible, but sufficing for accurate measurements. This applies to thermo-galvanism, respecting which many philosophers have the erroneous notion that it cannot pass through a very long chain. With the arrangements at Göttingen, and on the application of a thermo-galvanic apparatus of peculiar construction, the effect is produced merely by touching the connecting points with the finger.

The application to the common electricity of friction gives rise to another interesting observation. It is known that Colladon discovered, by experiments which were at first doubted, but subsequently confirmed by Faraday, that the common electricity of friction, conducted through a multiplier, deflects the needle in the same manner as a hydro-galvanic current. Faraday was the first to prove that, in a very powerful electrical battery, no more electricity is developed than very weak hydro-galvanic means of excitation propel in a few seconds through a conducting wire of moderate length. Both the reality and the small amount of the electromagnetic action of machine electricity were experimentally confirmed several years ago with the apparatus here; it appeared, however, worth while to repeat these experiments with the aid of the new and much more sensitive apparatus. Instead of discharging a Leyden jar, or a battery of jars, by a wire chain (as Colladon and Faraday did), only the conductor and the rubber of an electric machine in the Physical Cabinet were connected with the wire chain passing to the Astronomical Observatory, which was 13,000 feet in length, including the multiplier. The electrical machine was then turned with uniform velocity; when this was done with a velocity of one revolution to a second, the five-and-twenty-pound magnet bar in the new apparatus in the Astronomical Observatory was thereby kept at a deflection corresponding to 144 parts of the scale (somewhat more than 50 minutes), — the deflection being positive or negative according to the direction in which the current passed through the multiplier. The experiments showed as much regularity as could be wished. But the circumstance especially remarkable is that the electromagnetic effect remained the same even when the length of the chain was increased to above a German mile by the introduction of other apparatus. This might seem to be an essential difference from other currents, excited either

[251][Note by AKTA:] In German: *Kette*. This word may also be translated as "circuit" or "electric circuit".

hydro-galvanically, thermo-galvanically, or by induction; the intensity of which, indicated by the magnitude of the electro-magnetic effects, becomes constantly smaller the longer the conducting apparatus; but I find in it a striking confirmation of the theory, according to which the unequal intensity, indicated by the unequal electro-magnetic action of two galvanic currents, is nothing more than the unequal quantity of electricity passing through each section of the conducting apparatus in a fixed time. With other modes of excitation, a given electromotive force developes less electricity in a given time, the greater the opposition made to the current by the longer chain; in our experiment, on the contrary, the quantity of electricity in motion depends merely on the play of the machine, and all electricity passing to the conductor in the form of sparks must traverse the whole chain, be it long or short, in order to equalize itself with the opposite electricity from the rubber.

In order to demonstrate the preference due to the new apparatus over the magnetometer in electro-magnetic telegraphy, we must first consider somewhat more closely *the manner of producing* telegraphic signs, by means of galvanic currents.

As soon as it was known that the action of a voltaic pile propagated itself through a very long chain, the inference seemed obvious that these natural forces might be employed for telegraphic purposes; and, thirty years ago,[252] when but a small portion of the galvanic actions were known, Sömmering proposed the evolution of gas for the purpose. The magnetic actions of galvanic currents, which were subsequently discovered, are far better adapted for complicated signals; yet it is surprising, that since Oersted's discovery,[253] many years elapsed before any one seems to have thought of applying them to this purpose. It is true that it was not possible to form a well grounded opinion of their applicability on a large scale, without an accurate quantitative knowledge of the decrease in force of galvanic currents, resulting from the length and quality of the conducting wires; on which subject, before Ohm and Fechner, very imperfect and erroneous notions were entertained. With a view chiefly to the performance, on a large scale, of similar experiments on the law of the force of galvanic currents under different circumstances, a connecting wire was established in 1833 between the Astronomical Observatory and the Physical Cabinet; the merit of the execution of this difficult project is due only to Prof. Weber.[254] This chain was from the very beginning frequently employed for telegraphic signals; not merely for simple ones in the daily comparison of the clocks, but complicated signals were also tried for the sake of experiment, and the possibility of communicating letters, words, and whole phrases, was even

[252][Note by CFG:] I have learnt from a note communicated to me by von Humboldt, that Bétancourt had, ten years before, laid down a wire from Aranjuez to Madrid, by means of which telegraphic signals could be effected by the discharge of a Leyden flask. Although no detailed account of the result appears to be known, there can be no doubt of the success of such an experiment, if properly performed; but such a method must always have been limited to conveying an affirmative or negative reply to a few previously-concerted questions.

[253][Note by AKTA:] Hans Christin Ørsted (1777-1851). See [Oer20b], [Oer20a], [Oer20c], [Oer65], [Ørs86] and [Ørs98]. See also [Fra81] and [Rei13].

[254][Note by AKTA:] Wilhelm Weber and Gauss invented in 1833 the world's first operational electromagnetic telegraph, [LB67, Section 66: Gauss and Weber's telegraph, pp. 41-42], [Ano89], [Fey33a], [Fey33b], [Wie60, Chapter 5, pp. 17-20], [Wie67, pp. 85-90], [Tim05], [Wol05], [MRGL10] and `https://www.uni-goettingen.de/de/historische-sammlung/47114.html`. It was a 3 km long twin lead connecting Göttingen University, where Weber was Professor of Physics, with the Astronomical Observatory (*Sternwarte*), directed by Gauss. This telegraph worked based on Faraday's law of induction discovered two years earlier, [Far32a] with German translation in [Far32b] and [Far89], and with Portuguese translation in [Far11]:

then an ascertained fact.[255,256] In these experiments an hydro-galvanic current was employed, excited only by very weak means, viz. a single or a double pair of plates, and unacidulated water; I shall not, however, stop here to describe the method then employed, as I have since substituted for it one entirely different. In the first method there was this inconvenience; that with our simple chain and the arrangement of the apparatus then adopted, (such experiments being merely a subordinate object) no more than two letters could be signaled in a minute. Even with the new arrangement expressly formed for the purpose, this velocity (which is obviously unconnected with the length of the chain, or the distance apart of its extremities) could not be considerably increased as long as only a simple chain was employed, though it would be increased to a very high degree with a compound one; but there was no sufficient reason in this case for establishing a chain of the latter kind, as there could be no doubt as to the result, and its real scientific value would have borne no proportion to the expense.

But the laws of induction have led me to a quite different method, in which a simple chain has been employed for more than two years, with complete success, for a much more rapid telegraphing. It will be the more allowable to dwell longer on this subject, as hitherto I have published no details concerning it. I have elsewhere described,[257,258] many years ago, the apparatus which I term inductor. I must however remark, that instead of the inductor

Friedrich Gauß und Wilhelm Weber beim Arbeiten mit dem von ihnen erfundenen ersten elektrischen Nadel-Telegraphen
Links Weber am Gebe-Apparat, an dem er die stromweckende Spule bewegt; rechts Gauß am Empfänger, durch das Fernrohr auf den Spiegel des Galvanometers schauend. In der Multiplikatorspule unter dem Spiegel die Magnetnadel, ein schwerer Eisenstab. Die Verbindungsdrähte sind über die Stadt Göttingen gespannt und auch am Johannisturm befestigt. Nach einer Zeichnung von A. Wald.
(Zu Seite 18)

[255][Note by CFG:] The first public notice of these experiments is in the *Gött. gelehrten Anzeige*, 1834, Aug. 9, p. 1273. See Schumacher's *Jahrbuch* for 1836, p. 38.

[256][Note by AKTA:] [Gau34a] and [Gau36b, p. 38 of Schmacher's *Jahrbuch* and p. 339 of Gauss' *Werke*].

[257][Note by CFG:] *Gött. gelehrten Anzeige*, 1835, p. 351; Schumacher's *Jahrbuch* for 1836, p. 41.

[258][Note by AKTA:] [Gau35b, p. 351] and [Gau36b, p. 41 of Schumacher's *Jahrbuch* and p. 341 of Gauss' *Werke*].

of 1050 coils described in the first notice, and of that subsequently increased to 3537 coils, the present one consists of 7000 coils, — the length of the wire alone amounting to more than 7000 feet. By a very simple manipulation with this inductor, (viz. by removing it quickly from a double magnet bar, on which it is at first placed, and then bringing it back immediately to its former position, without reversing it,) two powerful opposite galvanic currents are caused to pass through the conducting wire, one quickly after the other, and each lasting only an extremely short time. The effect of these two currents upon a magnet bar surrounded by a multiplier, and situated anywhere in the chain, consists in this: that it produces for a moment a very quick velocity, which is immediately destroyed. The needle, therefore, makes a very rapid but small movement either to the right or to the left, according as may be desired, and then is immediately at rest. It is evident that the changes of such rapid movements may be combined in various ways, and may be employed for signaling letters. Some degree of practice will of course be required to give the signals rapidly and precisely on the one hand, and on the other to read them with ease and certainty; but even by unpractised persons, about seven letters can easily be signaled in one minute, as many experiments have shown. If, instead of manipulation, appropriate mechanical arrangements were adopted, the velocity and precision would undoubtedly be considerably increased.

It is precisely in this kind of telegraphing that the new apparatus possesses a considerable advantage over the magnetometer; and for the following reasons. Although the two opposite impulses, of which one simple signal consists, are exactly *equal* in *force*, — and consequently the second destroys just as much velocity as the first produced, — yet the needle cannot be in perfect quiescence between the signals, because this perfect quiescence is only possible when the needle is in its natural position of equilibrium. Even if it is in this position *previous* to a signal, it is somewhat disturbed therefrom by the signal itself, and the directive force acting on the needle causes it to tend to return. Though a single signal causes only a very slight movement, yet a considerable disturbance from the natural position of equilibrium will arise from the accumulation of a great number of signals; and the result will be so much motion between the signals that they will lose somewhat of their sharpness of expression. It will easily be seen, on consideration, that under circumstances otherwise similar, this disadvantage is greater when the needle employed has a short interval of vibration, than when it has a long one. Its effect is greater, therefore, on the magnetometer in the Magnetic Observatory, than on the 25-lb. needle suspended in the Astronomical Observatory; and is least of all on the new apparatus, when its magnet bar, by being placed in the reverse position, is converted almost into an astatic needle. Thus, even when the needle is at a considerable distance from its position of equilibrium, the comparatively weak directive force, with which it tends to return to that position, does not produce in it any movements which can materially disturb the signals, while the current in the multiplier acts as strongly on the needle, and consequently produces quite as rapid movements, as if it belonged to an ordinary magnetometer.

A peculiar apparatus, which I have lately caused to be constructed, is highly useful in preventing the disadvantages and inconveniences arising from untimely vibratory motions, both in this kind of telegraphing, and in many other applications of magnetic apparatus. I give it the name of a *damper*, as its action consists in entirely destroying, in a very short time, vibratory motions, which would otherwise be continued for several hours. The damper constructed at first for the magnetometer in the magnetic observatory produced this effect in a very high degree, so that the greatest vibratory motions disappeared entirely in a few minutes. A similar arrangement can be applied to any vibrating needle, to the magnetometer, and to the new apparatus here treated of; and will certainly form an essential part of every apparatus which is to be employed for telegraphing by the method described above. A more complete explanation of this apparatus would, however, lead us too far from our present subject.

No particular name has been as yet given to the new apparatus. From its chief application it might be termed an *Intensitometer.* But as it is applicable to as many and as accurate magnetic measurements as the magnetometer, it has perhaps an equal claim to the name. The essential difference is, that the new apparatus is suspended by *two* threads, by which a new directive force is obtained with which the magnetic force is commensurable. The other differences, viz. in the mode of attaching the mirror, and in the means of measuring the relative amount of rotation of the several parts of the instrument, are conditions necessarily arising from the objects to be obtained. The new apparatus may therefore be termed a *bifilar* or *bipensil-magnetometer,* to distinguish it from the older instrument, the simple or *unifilar* magnetometer. I may express my conviction, that its more extended use, and especially its employment, conjointly with the declination magnetometer, in the term observations, at stations widely remote from each other, will be soon followed by an important progress in our knowledge of the disturbances of the earth's magnetism.

Chapter 12

[Weber, 1838] Observations on the Arrangement and Use of the Bifilar Magnetometer

Wilhelm Weber[259,260]

After the full development in the preceding article[261,262] of the principle of the *bifilar magnetometer*, and of all that is essentially necessary for its construction and application, an exact drawing of the instrument will be particularly interesting. The drawing (Plate on pages 136 and 149) is so accurate that any skilful artist can work from it. The following observations are added with a view of rendering the drawing still more intelligible, and of facilitating the adjustment of the instrument by other observers, as far as may be done by such directions.

259[Web38a] with English translation in [Web41d] and [Web66b].

260The Notes by E. Riecke, the editor of the second volume of Weber's *Werke*, are represented by [Note by ER:]; the Notes by Wilhelm Weber are represented by [Note by WW:]; while the Notes by A. K. T. Assis are represented by [Note by AKTA:].

261[Note by ER:] Ueber ein neues, zunächst zur unmittelbaren Beobachtung der Veränderungen in der Intensität des horizontalen Theiles des Erdmagnetismus bestimmtes Instrument. *Resultate*, 1837, I, p. 1. Gauss' *Werke*, Vol. V, p. 357.

262[Note by AKTA:] [Gau38b] with English translation in [Gau41c].

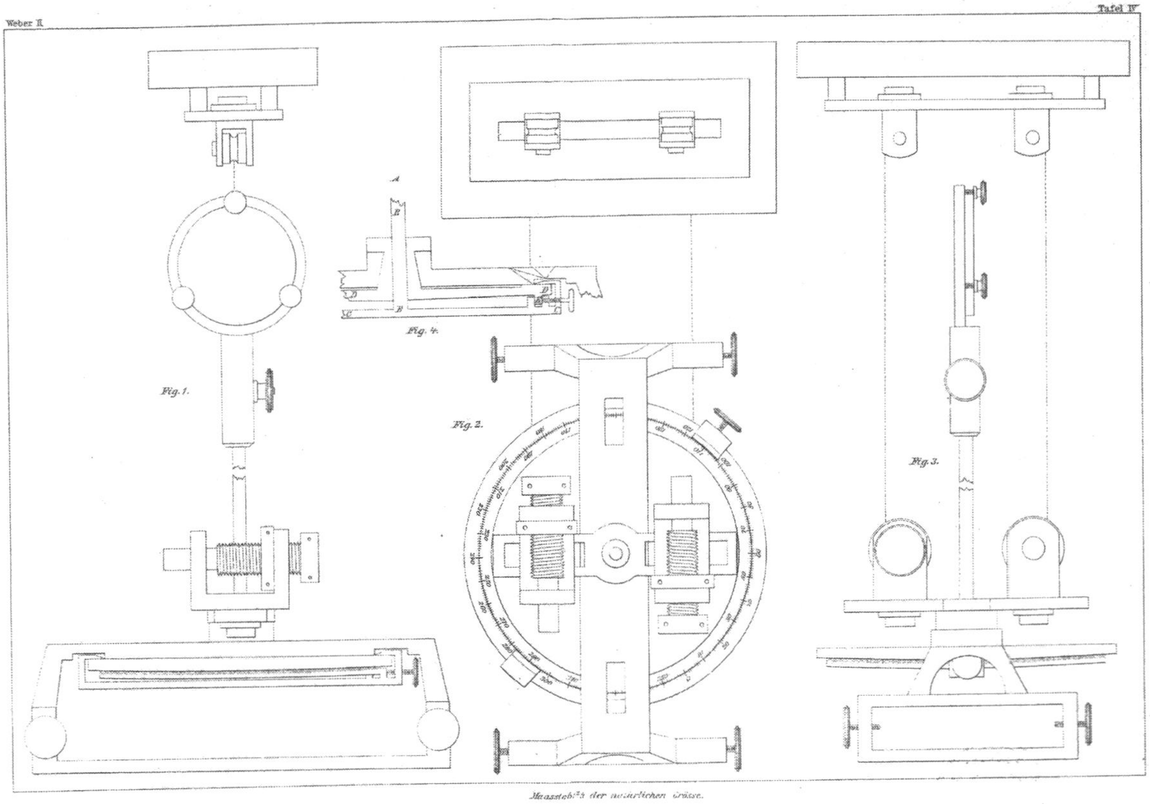

12.1 General Observations

The height and other dimensions of the Göttingen Astronomical Observatory, where the instrument figured has been established, allowed of large size in the instrument, and therefore a 25 pound bar very powerfully magnetized was employed. At other places it will perhaps be necessary to employ smaller dimensions, and we shall notice at the conclusion the difference in the cost produced by diminishing the size. Large dimensions are generally, however, more to be recommended for the bifilar magnetometer than for the unifilar, for the following reasons: 1st, because no proportionate increase in the price is occasioned, as the principal expense arises from the fine division of the circle and from the mirror; and since the latter is not attached to the end of the magnet bar, it is not requisite that its size should be increased with that of the bar; 2nd, because the enlargement of the instrument does not require any considerable enlargement of the room; which would be the case with the unifilar magnetometer, on account of the experiments of deflection in the measurements of the absolute intensity; 3rd, because the magnet bar need very rarely be removed from the stirrup; and therefore the size of the bar produces no inconvenience in its use, which would be the case to a certain extent with the other magnetometer. It does not follow that a bar of exactly five-and-twenty-pounds' weight, such as we have used, must be employed; one of ten pounds will suffice for the most delicate measurements, and even one of four pounds might answer. The small bars have only one advantage over larger ones, in the greater facility of imparting to them a strong degree of magnetism; and this is only of importance where powerful means of exciting magnetism by friction are wanting.

With respect to a suitable locality, a room similar to that employed for the unifilar magnetometer is all that is requisite, even if a bar of 25 pounds is employed. The breadth of the room may even be less, and its length may form any angle with the magnetic meridian, because the mirror in this case is not attached to the extremity of the magnetic bar, but to the stirrup at the centre of the bar, and it can be turned in any direction. A considerable height is requisite, so that the interval of the two threads or wires to which the instrument is suspended may be sufficient for convenient measurement without rendering the directive force[263] too great. As it is rare that a room is sufficiently high, it is advisable to pierce the ceiling, and to carry the wires as high as the roof will allow. In regard to the height, it is of little consequence whether a heavy or light bar be placed in the stirrup, supposing only both bars to be proportionally magnetized, and both to be much heavier than the stirrup. It is not necessary to construct a separate building free from iron for the bifilar magnetometer, as is done for the other magnetometer; it may be placed, as is the case at Göttingen, in the middle of a room in a building from which iron has not been excluded: it is sufficient to remove all iron from the immediate neighbourhood of the instrument: it is best, however, to place it in the magnetic observatory which contains the other magnetometer, if the room is large enough and adapted for the purpose. If, for instance, the changes of the declination and of the intensity are to be observed *simultaneously* during the terms, a double number of observers is necessary if the apparatus are in different buildings. But if both are in *one* large room, and so arranged that, whilst the magnetometers are at a sufficient distance asunder, the theodolites with which the observations are made are situated near one another, one clock may serve both observers, and one practised observer may observe alternately with both instruments, allowing an interval of two minutes. The two magnetometers may be so placed relatively to each other in a large room, that the mean declination may remain unaltered, and the changes of the declination and of the intensity be only so far affected, that the determination of the value of the divisions of the scale is somewhat different from what it would otherwise have been. This is the case when the pillar supporting the theodolites forms with the two magnetometers a triangle, of which one side (viz. that between the pillar and the declination-magnetometer) is situated in the magnetic meridian, while the other side, viz. the line which connects the central points of the two magnetometers, forms an angle of $35°15'52''$ with the magnetic meridian.[264]

---

[263] [Note by AKTA:] The concept of "Direktionskraft" (directive force) was introduced by Gauss in 1838, [Gau38b, p. 4] with English translation in [Gau41c, p. 254], see footnote 246 on page 125. Alternative translations: force of direction, directing force or directional force

[264] [Note by WW:] Prof. Gauss has given, in a very simple geometrical construction, the complete solution of the problem of the reciprocal action of two magnets at a great distance, in any given position relatively to each other. It is as follows:

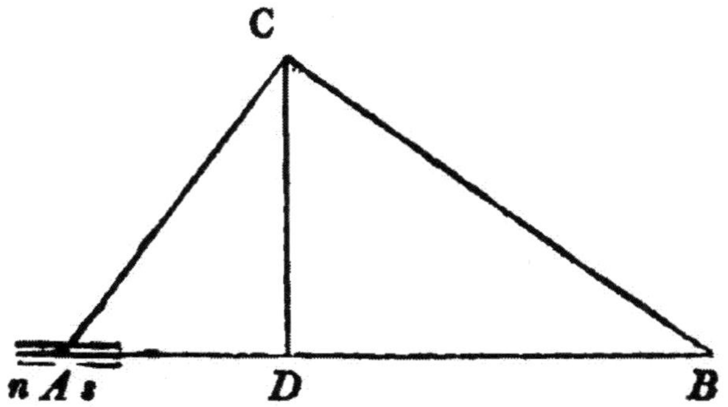

Let A be the centre of a small magnet, ns; AB the prolongation of ns; C a particle of free magnetism of the other bar; ACB a right angle; $AD = \frac{1}{3}AB$; then CD *is the direction of the force* which acts upon C, when C is a north magnetic particle; (when C is a south magnetic particle, the direction of the force is, on

The height of suspension, which is of such great importance for the objects of this instrument, renders it very desirable that access to the points of suspension should be rarely or never required. Even in the construction of the unifilar magnetometer it was noticed that it would be convenient that the torsion circle, of which frequent use is made, should be fixed to the stirrup of the magnetometer instead of to the ceiling. The same object has been considered in the construction of the bifilar magnetometer, where, on account of the

---

the contrary, in the prolongation of DC beyond C) $\frac{CD}{AD} \cdot \frac{Mm}{AC^3}$ is *the magnitude of the force*, M designating the magnetism of ns, and m the magnetism at C. This simple proposition, which is useful in numberless cases, is especially applicable to this case, in which the most advantageous reciprocal position of the magnetometers to be placed in the same room is required; i.e. that position in which they will least disturb each other, and in which, whatever slight disturbance may be produced can easily be brought into calculation as a correction. The application of Gauss's proposition to our case shows that in the position above described, 1st, the *mean* declination remains unchanged; 2nd, the value of the divisions of the scale, not only for the *variations* of the declination, but also of the intensity, are only altered in so far as the *directive force* of the two apparatus undergoes a change; for the value of the divisions of the scale changes with the directive force, and in the same proportion. This may all be seen from the geometrical construction of the reciprocal action of two magnets at a great distance, without its being necessary to give a detailed development of *the theory* of the two magnetometers.

The *first* assertion is evident from the consideration of the above Figure, where A is the central point of the intensity-bar ns, C the central point of the declination-bar situated in the line CD, CD the magnetic meridian, and where the straight line AC, which connects the centres of the two bars, forms the angle $ACD = 35°15'52''$ with the magnetic meridian CD, — or, more accurately, forms such an angle, ACD, that

$$\sin ACD = \sqrt{\frac{1}{3}} \, ,$$

$$\cot ACD = \sqrt{2} \, ,$$

$$\csc ACD = \sqrt{3} \, .$$

According to the above proposition, CD is the *direction* of the force which acts on the declination-bar C, for if $ACB = 90°$, $AD = \frac{1}{3}AB$. This latter case is the actual one, because CD is perpendicular to AB (the magnetic axis of the declination-bar must be situated in the magnetic meridian, and the magnetic axis AB, of the intensity-bar must be placed perpendicularly to it); then AC being the half diameter, AD is the sine of ACD, AB the secant of BAC, or the cosecant of ACD; consequently,

$$AD : AB = \sin ACD : \csc ACD = \sqrt{\frac{1}{3}} : \sqrt{3} = 1 : 3 \, .$$

The direction of the force with which the intensity-bar acts on the declination bar is therefore that of the magnetic meridian CD: it may consequently have some influence on the time of vibration of the declination-bar, the directive force of which is somewhat changed by it; but it will exert no influence on its direction, so long as this direction coincides with the assumed *mean* meridian CD: the *deviations* from it will, however, be somewhat diminished or increased by this force, according as it acts conjointly with, or in opposition to, the terrestrial magnetic force; but even this is provided for if we alter the value in arc of the divisions of the scale, in which the *deviations from the mean meridian* are expressed proportionately to the force of direction, i.e. by

$$\frac{\sqrt{2}}{AC^3} \cdot \frac{Mm}{Tm} \, ,$$

where, according to the above proposition,

$$\frac{Mm}{AC^3} \cdot \sqrt{2} = \frac{Mm}{AC^3} \cdot \cot ACD = \frac{CD}{AD} \cdot \frac{Mm}{AC^3} \, ,$$

designates the *magnitude* of the directive force produced by the intensity-bar, and Tm designates the directive force of the earth.

greater height of suspension, it was of much more consequence. To make it quite unnecessary to go to the ceiling in the case of the bifilar magnetometer, several other arrangements are requisite at the stirrup. The screws, for instance, which serve to lengthen and shorten the wires, have to be fixed to the stirrup instead of the ceiling. They are very clearly marked in Figures 1, 2 and 3 of the Plate, where it is seen how they are connected with the circle on which the stirrup is placed, and arranged in the same way as the elevating screw of the unifilar magnetometer, so that the wires may be lengthened or shortened without any lateral displacement.

The *latter* assertion, in so far as it relates to the changes of the intensity, is proved by letting fall a perpendicular on CA at C, which intersects the prolongation of the line CD at E.

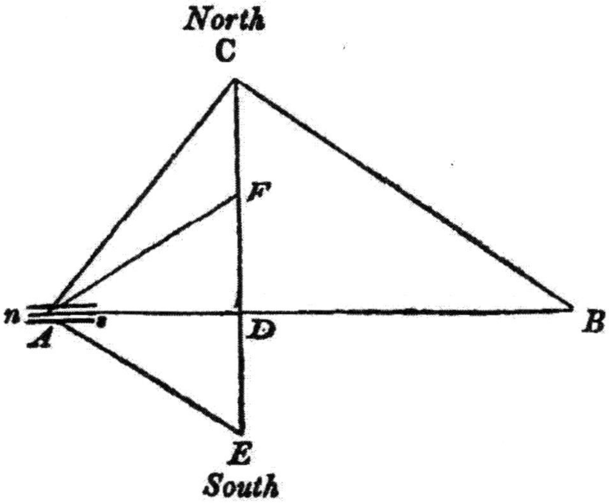

It then results from the similarity of the triangles ACD, ABC, EAD, ECA, that $ED = \frac{1}{3}EC$, because AD was equal to $\frac{1}{3}AB$; consequently, if CD is bisected in F, $CF = \frac{1}{3}CE$. Now, as all that has been said of A, AB, C, ABC, AD, and CD, is true also of C, CE, A, CAE, CF, and AF, it results that AF is the *direction*, and $\frac{AF}{CF} \cdot \frac{Mm}{AC^3}$ is the *magnitude* of the force with which the declination-bar acts on the intensity-bar. If now we resolve this force into a force acting in its magnetic axis, by multiplying the entire force by the fraction AD/AF, and into a force perpendicular to it (towards the magnetic meridian) by multiplying the entire force by the fraction DF/AF, we obtain for the *first* the value

$$\frac{AD}{AF} \cdot \frac{AF}{CF} \cdot \frac{Mm}{AC^3} = \frac{AD}{CF} \cdot \frac{Mm}{AC^3} = \frac{Mm}{AC^3} \cdot \sqrt{2}\ ;$$

and for the *latter* the value

$$\frac{DF}{AF} \cdot \frac{AF}{CF} \cdot \frac{Mm}{AC^3} = \frac{DF}{CF} \cdot \frac{Mm}{AC^3} = \frac{Mm}{AC^3}\ ,$$

since $2CF = CD = AD\sqrt{2}$, or $AD/CF = \sqrt{2}$, and $CF = DF$.

The force $\frac{Mm}{AC^3}\sqrt{2}$ directly changes the *directive force* of the intensity-bar in its transversal position. This force Mm/AC^3 would alter its *position* if the effect were not counteracted by a suitable change in the suspension, so that the bar should remain unmoved in its transversal position. In the latter case it is true that the force Mm/AC^3 no longer comes into consideration; but the changed suspension has certainly some influence on the directive force, and consequently on the value of the divisions of the scale. This, however, does not require any separate calculation, being included in the calculation of the directive force from the given suspension, — a problem which belongs to the *Theory of the Bifilar Magnetometer*, which will subsequently be developed.

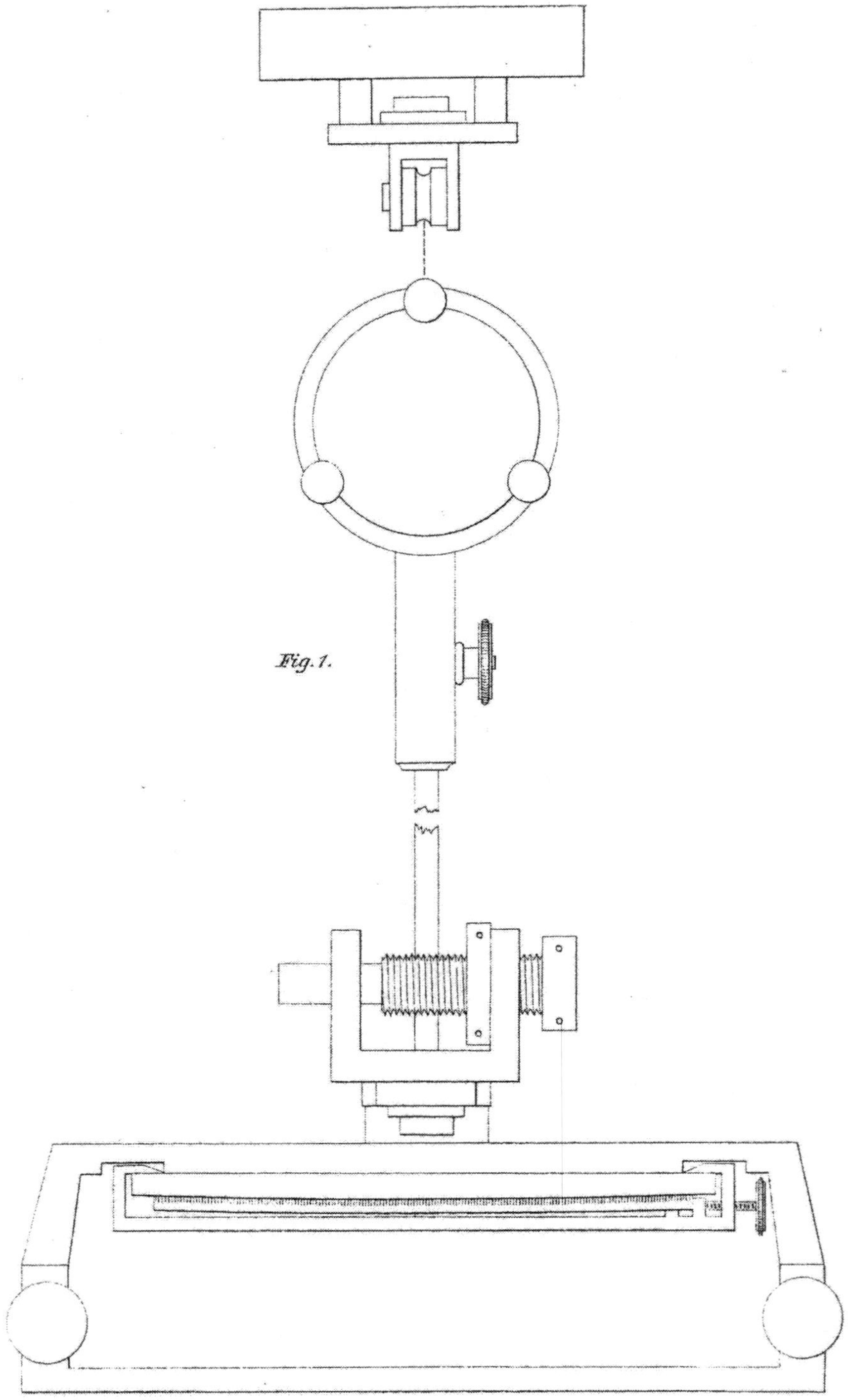

Fig. 1.

Fig. 2.

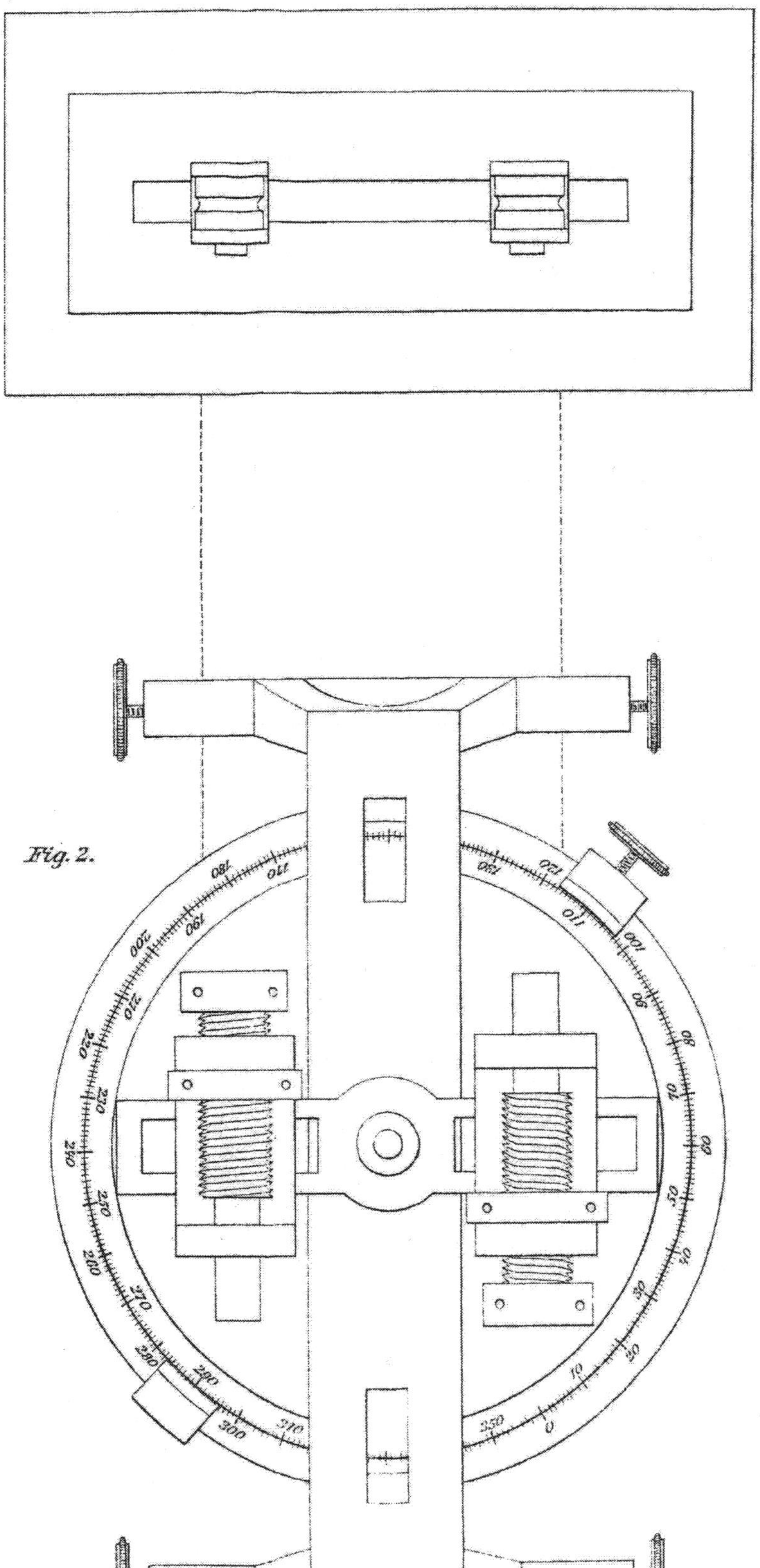

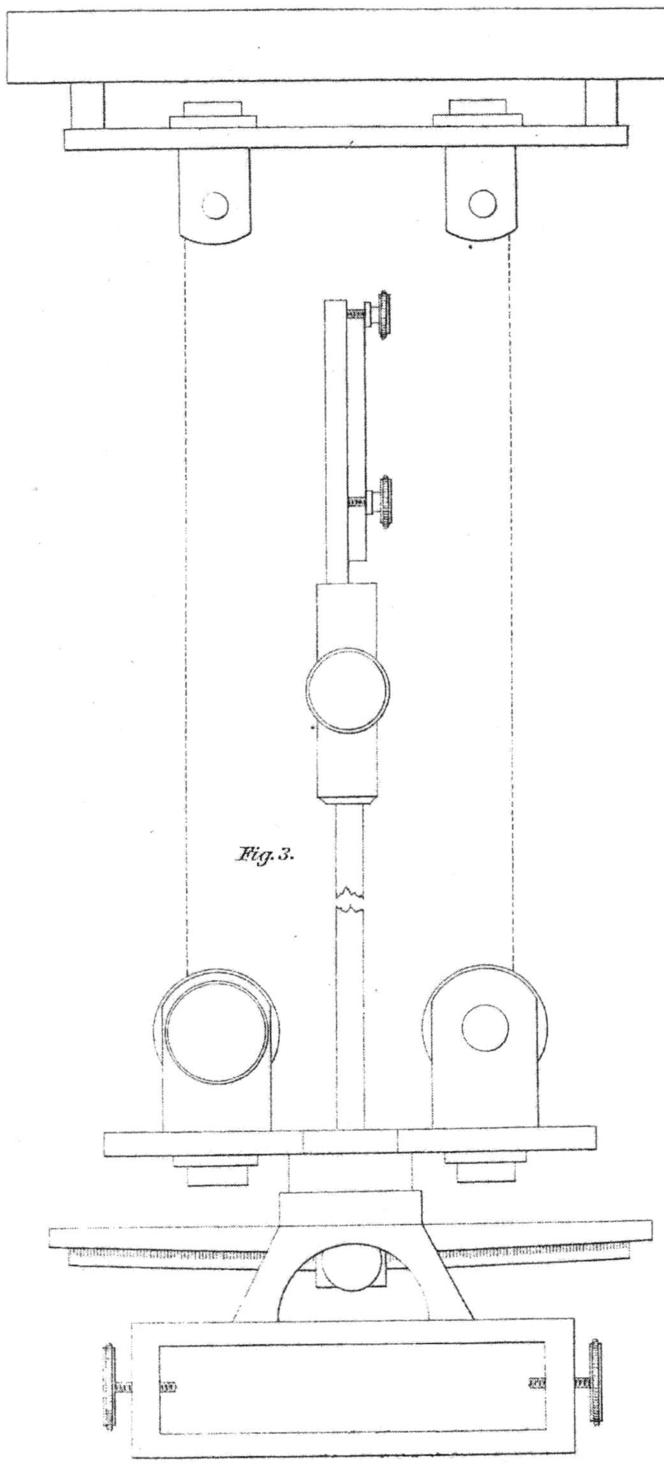

It is also necessary to be able to bring the two wires closer to, or further from each other, at the stirrup, so as to increase or diminish their directive force at pleasure. Although it is most simple that the two wires which support the instrument should be always equi-distant above and below, and that whenever it is desired to increase or diminish the directive force, they should be moved through an equal quantity at both extremities, it is by no means necessary. The change in the interval of the wires may be effected below only, but in such case to a greater degree. The apparatus figured is, in fact, so arranged, that, with a mean distance of the upper ends, every necessary increase or diminution of the directive force can be produced by a displacement of the suspension screws at the stirrup; however, for the sake of completeness, the apparatus is provided with an arrangement at top for an equal displacement of the two cylinders, over which the wire is conducted, and by which its two vertical suspended ends are kept separate from each other; so that, if it is desired, the upper distance may always be rendered equal to the lower. In case it is not desired to retain the power of making this upper displacement, these *two cylinders* may be united into *one roller* of a suitable diameter, and the axis of this roller, like that of a friction wheel, may be allowed to run on wheels, so as to diminish the friction, and cause the two wires to have an equal tension; — a point which is of great importance in absolute determinations.

12.2 On the Separate Parts of the Bifilar Magnetometer

The description of the individual parts of the bifilar magnetometer is reduced almost wholly to a description of the stirrup, because it unites nearly all the parts which in the unifilar magnetometer are distributed among the stirrup, the ceiling, and the extremity of the bar. It is also unnecessary to speak of the theodolite and its stand, the clock, the scale, or the mark, as all these have been treated of in the account of the former instrument in the *Resultaten* etc. from last year, pp. 14-19.[265,266] But as so many arrangements are united in the stirrup, its construction requires to be particularly explained. The Plate gives three different views of the instrument, of the natural size, and as arranged for the 25 pound bar; the small and compound parts have been represented in a separate section, so as to exhibit their interior mechanism.

It requires an attentive consideration on account of the many important parts compressed into so small a space at the stirrup: a clear comprehension of its mechanism will be obtained when we know the various concentric rotations which are performed at the stirrup, — the mode of checking and measuring these, — and their objects. The rotations are the following.

1. Of the mirror on its pivot; — the whole of the other portions of the instrument remaining unchanged.

2. Of the mirror, with its pivot and alidade, on the circle to which the suspension-screws of the wires are fixed, and on which the stirrup and its alidade rest.

3. Of the stirrup with its alidade, on the circle on which it rests.

   In order to complete the view of all the rotations, we may here add,

4. That of the two upper extremities of the wire around one another, i.e. around the same axis as that on which the other rotations take place.

The *first* rotation will be sufficiently intelligible from Figures 1 and 3 of the Plate. The arrangement is simple, because its amount does not require to be measured. Its object is

[265][Note by ER:] W. Weber's *Werke*, Vol. II, pp. 2-6.

[266][Note by AKTA:] [Web37a, pp. 14-19 of the *Resultate aus den Beobachtungen des magnetischen Vereins im Jahre 1836* and pp. 2-6 of Weber's *Werke*] with English translation in [Web41f] and [GW39a]; and with French translation in [Web38c].

merely to allow of perfect freedom in fixing the theodolite; the axis of the mirror can always be made to revolve, to suit the position of the telescope and the scale, wherever they may be placed. The image of the scale which appears in the mirror serves itself to regulate the rotation, and no further arrangement for measuring it is required. A screw, as exhibited in the Figures 1 and 3, fixes the mirror in its position.

For the *second* rotation, the three pieces, the mirror, the pivot, and its alidade, are firmly connected as one piece, and revolve together in the cavity of the circle; they are represented, together with the latter, in a cross section, at Figure 4.

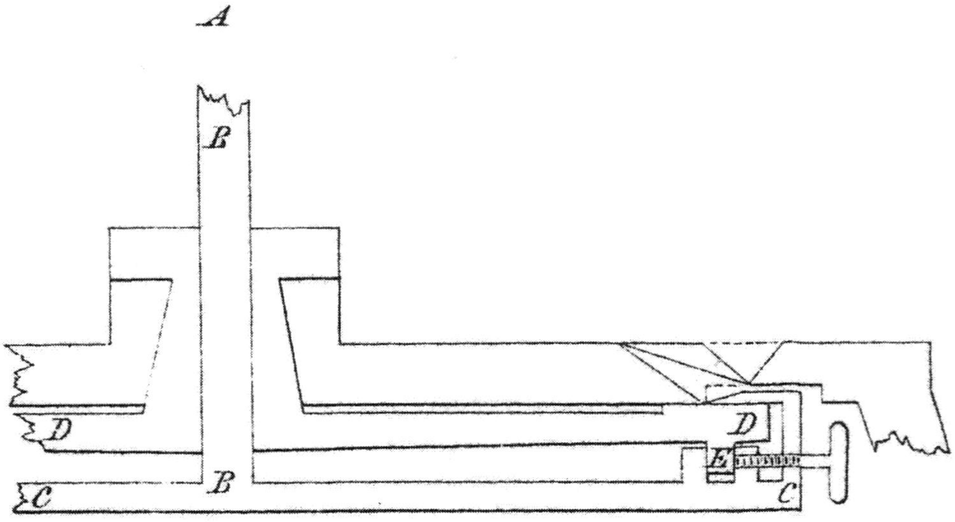

Fig. 4.

The mirror is placed on the upper end of the pivot B at A; C is the alidade of the pivot; D is the circle. The only essential difference between the second rotation and the first is that in the second the angular amount can be measured. As the revolving alidade of the pivot, situated beneath the circle, embraces at its two extremities the edge of the circle, it forms on its upper and graduated surface two verniers,[267] the inner margins of which lie close to the outer margin of the divided circle. A clamp, by which the alidade of the pivot can be pressed firmly against the circle, is seen in the section at E, Figure 4.

The second rotation alone would be sufficient if there were at no time an impediment to its use. The verniers on the alidade of the pivot come in certain cases beneath, and are hidden by the alidade of the stirrup. In the instrument represented in the Plate much care has been employed to restrict this within very narrow limits, as will be plainly perceived in Figure 2; but, in order to meet the rare cases in which it does still occur, without having to alter the position of the theodolite, both rotations may be employed at the same time, so as to free the indices without turning the mirror from the scale.

The *third* rotation is that of the stirrup with its alidade, on the circle upon which it rests. The directive force of the wires acts immediately on the circle to which the suspension screws are fixed: the directive force of magnetism acts immediately on the stirrup in which

[267][Note by AKTA:] In German: Nonien. "Nonius" is a measuring instrument used in navigation and astronomy. It was named in honour of its inventor, the Portuguese scientist Pedro Nunes (1502-1578), written in Latin as Petrus Nonius. The nonius was created as a system for taking finer measurements on circular instruments such as the astrolabe. The system was adapted into the Vernier scale by the French scientist Pierre Vernier (1580-1637).

the magnet-bar is placed. When, therefore, these two directive forces form an angle with each other, the two parts upon which they act will have a tendency to move in opposite directions. That no such displacement of the parts may occur, they are made to slide on each other with so much friction, that the two directive forces, when forming a large angle with each other, may not be able to overcome it. For a similar reason it was provided in the unifilar magnetometer that the alidade of the stirrup should be placed on the *outermost* margin of the circle, so that the friction produced by its pressure might act with the greatest leverage. The same has been done with the bifilar magnetometer, where this provision is much more essential and important, the forces which tend to displace the two parts being much more powerful. Further, we must be able to measure with great exactness this rotation, on which depends the angle which the two directive forces form with each other. The simplicity of construction of the bifilar magnetometer consists chiefly in this circumstance, that the same circle and graduation serve for measuring both the second and the third rotation. For this reason the alidade of the stirrup is also furnished with two noniuses.[268] The instrument consists, therefore, of a circle with two alidades, which may be used independently of each other. In order that this independent use may never cause the two alidades to interfere, the one is situated beneath, and the other above the circle. But since each alidade is provided with two noniuses, and all four are to move on the divided limb of the circle, which is its upper surface, the inferior alidade embraces the margin of the circle and forms noniuses which abut at the outer margin, whilst those of the upper alidade, in order not to come in conflict with those of the lower, abut on the inner margin. The noniuses of the upper alidade can thus pass by those of the lower one, and even an interval may exist between them, which, however, must be smaller than the length of the divisions on the circle. Thus the graduation of the circle serves two purposes, the one not interfering with the other, only it cannot serve both purposes at the same time, as the figures must be covered either by the noniuses of the one or of the other alidade, according as they are inside or outside of the graduation. For this reason the figures are placed alternately on the inner and on the outer side, as exhibited in the Plate, Figure 2.

The *fourth* rotation is that of the two upper ends of the wires. No mechanical arrangement is required for this rotation; but the bearer on the ceiling, by which the wires are carried and adjusted, is turned by the hand. As the bearer must be fixed to the ceiling, no use is ordinarily made of this rotation; but it is so arranged in the first instance as to be in the most convenient position for all purposes. That position may be regarded as most convenient in which the lower ends of the wire interfere least with the mirror which is situated between them. It will be evident that, in the various uses to which this instrument is applied, if the bearer is not moved, the lower ends of the wire are brought into various positions, while the mirror retains its position between them nearly unchanged, being always directed towards the scale. The two wires, for instance, will sometimes be in one vertical plane throughout their whole length; sometimes they perform part of a revolution round each other, and a vertical plane drawn through them will form with the former one an angle, which is, however, always less than 90 degrees. If it be now so arranged that in the first case the plane of the wires coincides with the vertical plane of the optical axis of the telescope, the one wire passes just as far from the mirror in front as the other does behind, and both wires are as far as possible from the mirror. If the instrument is then arranged for the other use, the wires are brought nearer to the mirror, but not so as to touch it, even if the mirror were larger

[268] [Note by AKTA:] In German: *Nonien.* Previously this expression had been translated as "verniers", see footnote 267.

than the intervening space, because the rotation does not amount to 90°. It is always less than 90°, because the directive force arising from the suspension must be greater than the magnetic directive force; hence the moments of rotation arising from the two forces will only equilibrate when the wires undergo a smaller rotation than the magnetic axis; and since the latter, in the transverse, must be 90° from its natural position, it follows that the rotation of the wires must be less than 90°.

12.3 On the Use of the Bifilar Magnetometer

I shall in conclusion briefly notice the series of experiments which must be performed in fixing and adjusting the apparatus.

1. The clock, the theodolite, and the scale are fixed, and a plumb-line dropped from the centre of the object-glass across the scale. The theodolite is to be leveled.

2. The telescope is directed to the opposite wall, on which there is a mark, serving to designate the terminal point of the optical axis. The scale is placed perpendicularly to the vertical plane of that axis.

3. A place is sought in the vertical plane of the optical axis for the mirror, the distances of which from the centre of the object-glass, and from the division of the scale across which the plumb-line is suspended, are, together, equal to the distance of the mark from the centre of the object-glass. The horizontal plane of this point must bisect the plumb-line from the centre of the object-glass. A plumb-line is let fall from the ceiling through this point.

4. The bearer is either fixed to the ceiling, or perpendicularly above a hole made through the ceiling, from 80 to 100 millimetres wide; so that the ends of a thread passed over it, and extended by small weights, pass freely through the aperture, and are both situated in the vertical plane of the optical axis of the telescope.

5. One end of a steel wire, sufficiently strong to carry half the weight of the instrument without danger of breaking, is fastened to one end of the thread, and drawn up to the bearer by drawing the other end of the thread down (care being taken that the wire and the thread should always be extended in a straight line); it is passed over the two cylinders of the bearer, and drawn down; the thread is then removed, and the two ends of the wire, weighted, are left to hang freely until they have assumed their natural position.

6. The two ends of the wire are cut off about 100 or 150 millimetres below the place where the magnetometer is to be suspended, and are fixed to the suspension screws. The stirrup thus carried is then, with the help of the screws, wound up into its proper position.

7. A box, sufficiently large to contain the magnet-bar, is placed underneath to protect the instrument in case the wires should break, and to prevent currents of air. This box is closed on all sides. Its lid consists of two halves, which fit close, and leave only *one* round aperture, through the centre of which the pivot passes; the upper end of the pivot carries the mirror, which must be above the box. The two wires having the mirror between them pass through the same aperture. This circular aperture is usually closed by two semicircular flaps, in which there are small slits for the pivot and the wires.

8. Before the magnet bar is laid in the stirrup, a weight of the same size, but unmagnetic, is placed therein, and the wires are suffered to arrange themselves in their natural position, in which both are in one vertical plane throughout their whole length. The alidade of the stirrup is then brought as exactly as possible into the mean magnetic meridian from which the changes of variation are to be measured. The other alidade on the pivot of the mirror

should be so fixed as to form a right angle with the alidade of the stirrup, in order that the noniuses may be far apart. The weight in the stirrup is moved until the mirror is situated exactly between the two wires, when the axis of the mirror should be very nearly horizontal. Employ the first rotation to direct the mirror towards the scale, without disturbing the alidade. If the scale does not appear in the telescope, it will be seen by the naked eye a little above or beneath, and may be brought into the field by the help of a light running weight placed on the stirrup. The first observation is then performed, and the position of the scale determined.

9. The time of vibration for determining the directive force of the wires may be observed before the magnet-bar is inserted, and again with a known increase of the moment of inertia. It is better, however, to perform this experiment somewhat later, when the distance of the wires from each other has been accurately adjusted, in case this distance has not been previously determined by calculation, and regulated accordingly.

10. The magnet-bar is then placed in the reverse position, (north towards the south) and the position of the scale again observed: this ought to agree with the observation (8.) If the two readings do not coincide, agreement must be attained by merely turning the stirrup with its alidade. The coincidence of the two readings proves that the magnetic axis of the bar is situated in the magnetic meridian. The less the directive force arising from the mode of suspension exceeds the magnetic directive force, (see 8.), the more delicate is this test, so that it may be impossible to obtain a *perfect* coincidence of the two readings; a difference of a few divisions of the scale may then be considered as unimportant. The influence of the hourly variations must be attended to, by making continued observations with a second apparatus of the same kind, or by making continued observations of the time of vibration with a common magnetometer.

ll. The time of vibration, t, is observed in this reverse position.

12. The magnet-bar is then placed in its natural position, (north towards the north,) by turning the stirrup with its alidade exactly 180 degrees; the time of vibration, τ, is again observed. Then the magnetic directive force, M, is to the directive force arising from the mode of suspension, S, in the ratio

$$M : S = t^2 - \tau^2 : t^2 + \tau^2 \ .$$

When this proportion deviates much from unity, the wires must be brought nearer to or moved further from each other, until the altered directive force of the wires exceeds but little the magnetic directive force; for instance, by about the tenth part of the latter, like it did after the previous treatise.[269,270] This is the case in the Göttingen magnetometer.

13. Seek the angle z, the sine of which is

$$\sin z = \frac{t^2 - \tau^2}{t^2 + \tau^2} \ .$$

Turn the alidade of the stirrup (say in the direction of the daily motion of the sun) $90° - z$, and turn the alidade of the pivot of the mirror in the opposite direction through the angle z. The equilibrium is then disturbed: the wires can no longer remain in their natural position, but must turn the circle to which they are fixed (and thus the whole instrument) exactly through the angle z, in the direction of the daily revolution of the sun. In this new position

[269][Note by ER:] *Resultate.* 1837. I, p. 8. Gauss' *Werke*, Vol. V, p. 363.
[270][Note by AKTA:] [Gau38b] with English translation in [Gau41c].

the equilibrium may be re-established, since the bar makes with its former position an angle $(90^\circ - z) + z = 90^\circ$, while the wires have only been turned through the angle z at their lower ends. It follows, thence, that if the wires were previously in their natural position, and if the magnetic axis of the bar was situated in the magnetic meridian, the opposite moments of rotation arising from the two forces M and S are to each other in the proportion

$$M \sin 90^\circ : S \sin z \ .$$

But as

$$M : S = t^2 - \tau^2 : t^2 + \tau^2 \ ,$$

$$\sin z = \frac{t^2 - \tau^2}{t^2 + \tau^2} \ ,$$

$$\sin 90^\circ = 1 \ ,$$

the equality of these opposite moments of rotation, or the equilibrium of the instrument in this position, is the result. Whether the true position of equilibrium coincide with the calculated one or not, is proved immediately by an observation of the scale, which ought to be the same as before. For the mirror has been turned (together with the whole apparatus) the angle z in the direction of the daily motion of the sun; but having been turned by its independent motion the same angle z in the opposite direction, it consequently retains its first position, and the point of the scale is unchanged.

14. If, however, the observation shows an alteration of the scale, it follows that the supposition in the first experiment — of the magnetic axis of the bar being in the magnetic meridian — was not accurately fulfilled. The amount of the error can be calculated, and the experiments repeated. This calculation will be still more accurate and certain, if a corresponding experiment has been previously made, proceeding precisely as described in (13.), only making all the rotations in the contrary direction.

15. When the required coincidence has been obtained, the magnetometer remains in its transverse position. Its time of vibration is then, according to a simple theorem, the geometrical mean between the times of vibration t and τ, and the observations of changes of intensity can be arranged like those of the changes of declination. The changes of intensity are obtained in divisions of the scale. If we desire to convert them into fractions of the entire intensity, these are obtained by multiplying the arc value of the scalar divisions (expressed in parts of the radius) by

$$\cot z = \frac{2t\tau}{t^2 - \tau^2} \ ;$$

for the value in arc of the parts of the scale, expressed in parts of the radius, gives immediately the changes of intensity in parts of the directive force, which, under the prescribed conditions, is $S \cos z$. If we divide this directive force by the whole intensity, i.e., by $S \sin z$, we obtain by multiplying the value of the arc by the quotient, $\cot z$ — the changes of intensity in fractions of the whole intensity.[271]

[271] [Note by AKTA:] A final Section of this paper with the costs of the bifilar magnetometers of various sizes, as supplied by Mr. Meyerstein of Göttingen, has not been translated.

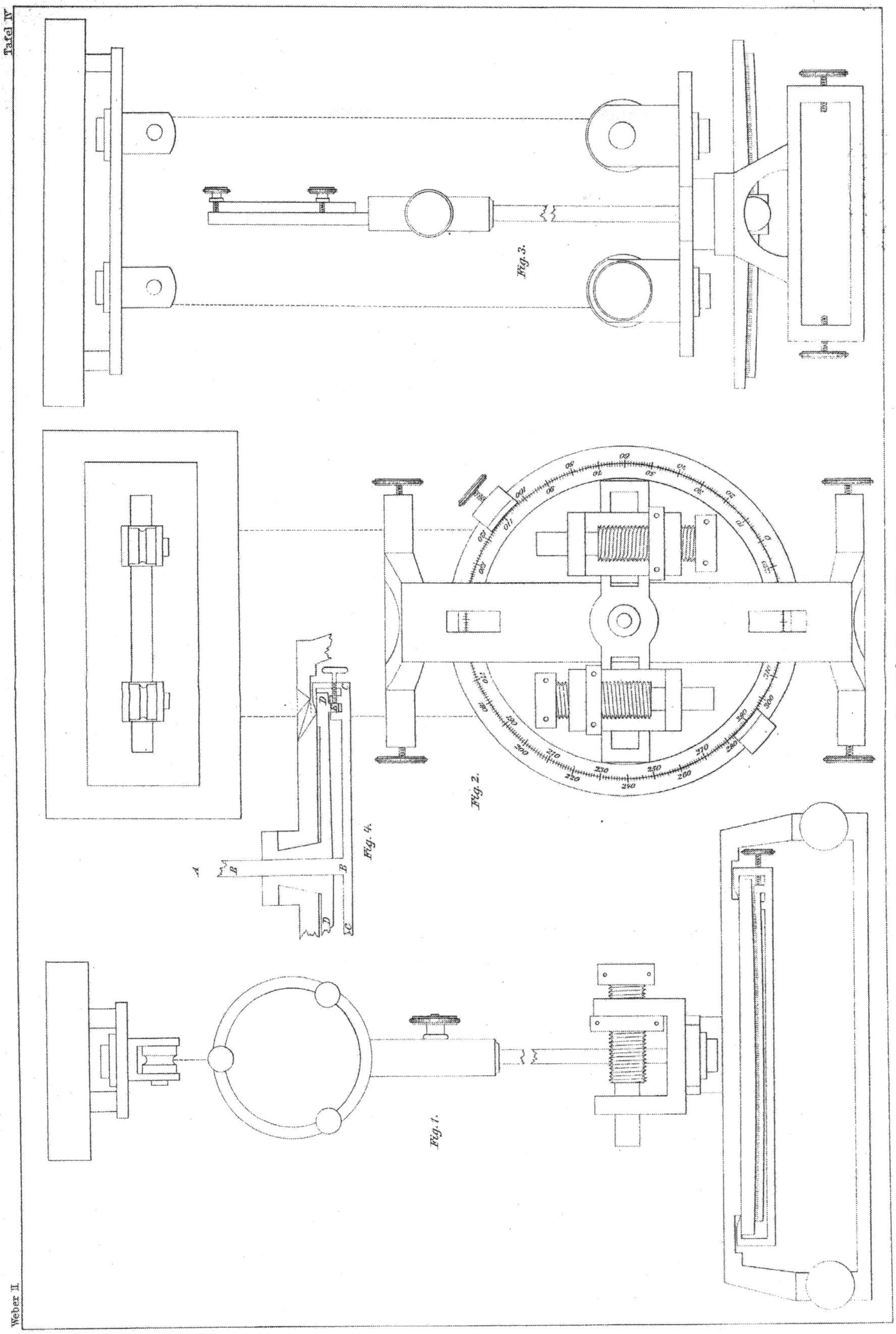

Maasstab: der natürlichen Grösse.

Chapter 13

[Weber, 1839] On a Transportable Magnetometer

Wilhelm Weber[272,273]

[This article is translated partly from the *Resultate aus den Beobachtungen des magnetischen Vereins im Jare 1838*, and partly from manuscript communications from M. Weber to Major Sabine.][274]

A small travelling apparatus for the absolute measurement of the force of the earth's magnetism has been described in the *Resultate* for 1836.[275,276] That apparatus was not a *magnetometer*, but rather served as an illustration of the mode in which this measurement, which had previously been executed only with a magnetometer, might be made with an ordinary *compass needle*.

The degree of accuracy attainable with such a small apparatus, and the occasions on which it ought to be employed, were examined in the memoir referred to. But for the limitation imposed by the want of time, or by other external circumstances, it would of course be always preferable to use the magnetometer; the small apparatus being only intended to serve as a substitute, on occasions when the use of the more perfect instrument is impracticable. It is very desirable to reduce the number of such occasions as much as possible, by devising means of removing the difficulties which often oppose themselves to the use of the magnetometer; and this will appear the more desirable, the more we consider the great difference in the degree of precision attainable by the two instruments; and the more we reflect on the importance that would be given to a class of observations in which magnetometers have not hitherto been used, (namely, those made during distant and extensive journeys and voyages,) if they could be rendered susceptible of a higher degree of accuracy, certainty, and completeness.

---

[272] [Web39] with English translation in [Web41e] and [Web66c].

[273] The Notes by Wilhelm Weber are represented by [Note by WW:]; the Notes by Eduard Riecke, the editor of the second volume of Weber's *Werke* are represented by [Note by ER:]; the Notes by Richard Taylor, the editor of the English translation of Weber's paper, are represented by [Note by RT:]; while the Notes by A. K. T. Assis are represented by [Note by AKTA:].

[274] [Note by AKTA:] This text appeared on page 565 of [Web41e], referring to Edward Sabine (1788-1883).

[275] [Note by RT:] Translated in the *Scientific Memoirs*, Part V.

[276] [Note by AKTA:] [Web37b] with English translation in [Web41a] and French translation in [Web38b].

If the final aim of such observations were simply that of constructing magnetic maps on which no ulterior investigation was to be based, the degree of exactness to which such maps should be carried might be arbitrarily determined; and possibly such an amount of accuracy as can be obtained without the use of magnetometers might be deemed sufficient. But if these maps are not themselves the final object sought, — if they are to form the basis of a new investigation, — if determinate rules and laws are to be recognised, — if the maps are to serve as the means of comparing experiment with the general theory of the earth's magnetism, — and if the elements of the theory are to be deduced from them, — then the degree of accuracy to be demanded is no longer arbitrary, but is determined by the nature of the subject. A *minor* degree of accuracy, such as these maps *now* possess, has, it is true, served for a first attempt at such a comparison; but in order that they may afford an adequate basis for an amended calculation, they must receive a *higher* degree of exactness. Such is now the *great purpose* of the magnetic observations to be made in distant expeditions, and it is this which now gives to such expeditions peculiar importance and value.

But the greater the importance which thus attaches to such voyages and observations, in consequence of the demands of theory, the more essential it becomes to examine what they are capable of affording.

Magnetic observations may be made at places widely remote from each other, either at the same time or nearly so, or alternately, so as to lessen the errors occasioned by regarding them as simultaneous. At all the stations, or at the more important at any rate, the observations may be continued with regularity for at least one or more weeks, so as to afford mean values freed in some measure from disturbing influences. But it is still more desirable to give to such expeditions the advantage of the recent improvements, by furnishing them with *magnetometers*. This would probably be best accomplished, by the persons who undertake magnetic expeditions making themselves thoroughly acquainted, both theoretically and practically, with the whole subject of magnetometric measurements, as they would then be able to devise for themselves the best travelling arrangements. But as there are not many opportunities of acquiring this knowledge, the following memoir may be interesting and useful to persons who cannot study the subject more thoroughly in other ways.

I proceed to describe a *transportable magnetometer*, which, as it unites all the advantages proper to magnetometers, with facility of management and compendious construction, appears well adapted for magnetic expeditions and journeys, and is not more inferior to the magnetometers of fixed observatories, than good portable astronomical instruments are to the larger ones used in fixed astronomical observatories. I shall *first* give some general remarks on this instrument; *then* a description of its several parts; and *lastly*, observations of the Declination, and its Variations, made simultaneously with the transportable magnetometer and with that of the Göttingen Observatory, and a measurement of the Intensity made for the purpose of exhibiting its capability in that respect.

13.1 General Remarks

The transportable magnetometer, figured in half size in Figure 1 of the Plate requires in general but few explanations, as it is only essentially distinguished from other magnetometers by its small size, and by its more compendious construction.[277]

[277][Note by AKTA:] A larger plate appears on page 181.

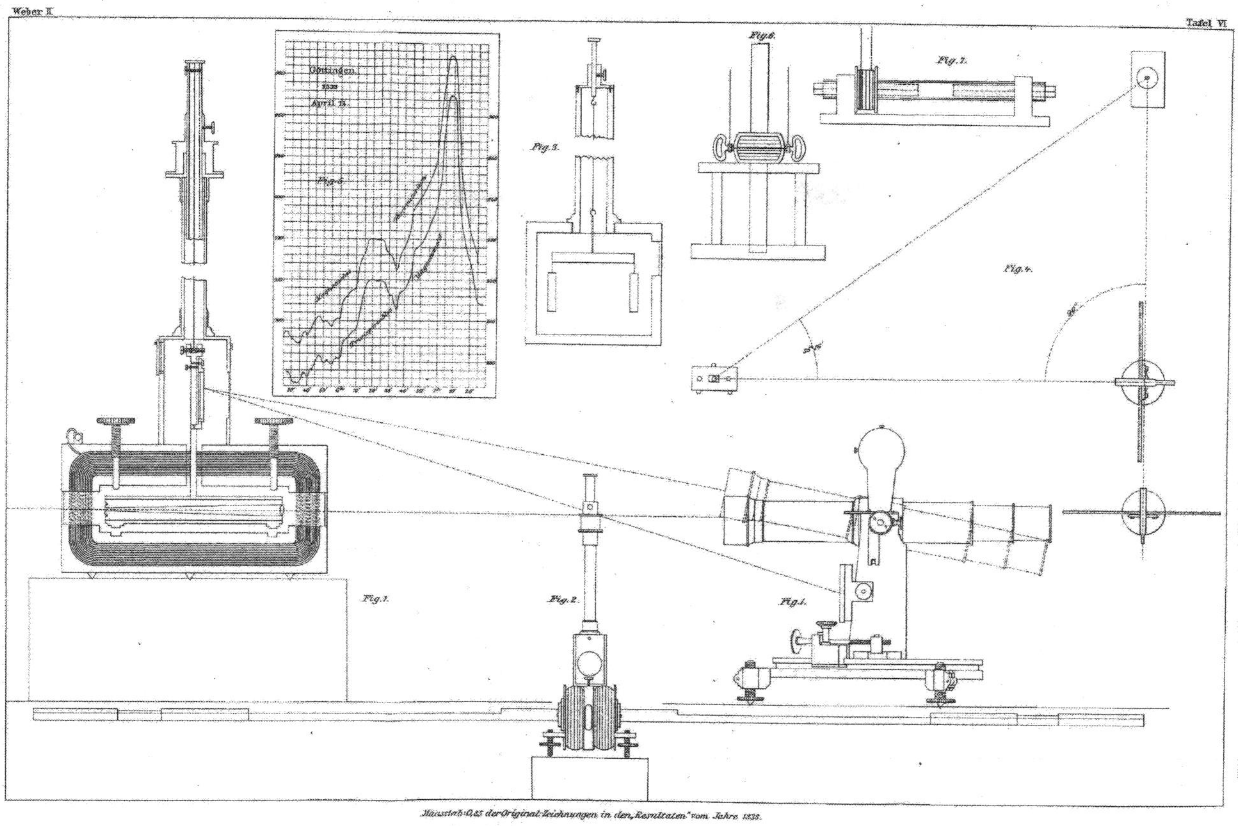

All the observations which are made with the larger magnetometers may also be made with the one under consideration; so that the *absolute declination*, the *variations of the declination*, and the *absolute horizontal intensity*, can all be measured by it. Since, like larger magnetometers, it is finally provided with a multiplier, all galvanic experiments can be made with it, and even, if a small rotation-inductor is used and the earth's magnetism is induced, the *absolute inclination measurement* can also be made.[278] The *variations of the horizontal intensity* can also be observed, by suspending the bar employed in the experiments of deflection, as a bifilar magnetometer. The exactness with which these various measurements can be made is much greater than has yet been attained in travelling observations; it suffices for all the purposes of magnetic travellers; and it admits of as much accuracy and certainty, in proportion to its size, as do the largest magnetometers.

The results obtained with the large instrument used in the Göttingen Magnetic Observatory may be depended upon almost to the immediate readings, which are to 1/10 of a division of the scale, or to 2 seconds of arc. This supposes the scale to be at least five meters from the mirror of the magnetometer, as otherwise the arc value of the divisions of the scale (which are one millimeter long), would be greater. Such a distance would not answer in journeys, as much time would be lost in bringing all the parts of the instrument into their proper positions. For travelling purposes, the distances ought to be limited so as to admit of the whole apparatus being placed on a table, and they should therefore be about four times less. Consequently, in lieu of the 8-inch theodolite, which is required to do full justice to the great magnetometers, one of about three or four inches may be used without disadvantage, being at once more convenient and more economical, and still allowing the measurements to be depended upon to within from 10 to 20 seconds of arc. In considering the subject further, it will be seen, that admitting the necessity in the travelling apparatus of diminishing the

[278][Note by AKTA:] This last sentence was not translated into [Web41e].

observation distance, a diminution in the size of the magnetometer (which would not be admissible under other circumstances), does in no degree detract from the accuracy of the observations. For with a distance four times less, the degree to which the reading can be depended on (and which it is desired to preserve), is not affected, though the proportion of the magnetic force of the magnetometer[279] to external disturbing influences be lessened in the same proportion. It may be assumed, that the magnetic force decreases as the cube of the linear dimensions of the bar, and external disturbing influences as the square, whence it follows that the bar may be made four times less without diminishing the dependence to be placed on the readings (which is to about the one tenth part of one division of the scale). If, with this diminution, other arrangements are adopted for guarding against external disturbing influences more carefully than has been hitherto found necessary with the larger magnetometers, there will be no material disadvantage in pushing the diminution in size somewhat further, having in such case only to preserve the degree of dependence which may be placed on the readings. In fact, the length of the bar of 600 millimeters has been reduced to 100 millimeters; and observation has shown that the readings may be equally depended upon; with this difference only, that the divisions, as read off, have a four times greater value of arc than in the case of the larger magnetometers, so that one division of the scale is equivalent to 80 seconds of arc instead of 20 seconds.

Hence it appears, that by suitable arrangements, all the advantages of the *magnetometer* may be secured to *magnetic expeditions*; of course, without that highest degree of precision attainable only in fixed observatories, where nothing is wanting in construction and arrangement.

The instrument to be described affords these advantages in respect to the *absolute declination* and its *variations*, and still more in respect to the *absolute measurement of the horizontal intensity*; for in the *Resultate* for 1836, p. 88,[280,281] it has been shown, that if both bars are six times smaller, the deflecting bar may be brought six times nearer to the magnetometer, without its being necessary to take more exactly into account the distribution of free magnetism in the bars. If, then, the length and breadth be diminished, and the thickness be left unaltered, (the large bars are 600 mm long, 36 mm broad, and 9 mm thick; and the small bars 100 mm long, 9 mm broad, and 9 mm thick,) it follows that as much may be gained in the small magnetometer, by increasing the angular deflection, as is lost by diminishing the distance of observation. In fact, the *experiments of deflection* admit of a precision which leaves nothing to be desired, and which harmonizes perfectly with the degree of accuracy which is known to be of easy attainment in the *experiments of vibration*.

Of course the small magnetometer must be constructed in such a manner that all its parts may form a solid whole, so that their relative position may not be liable to be disarranged by packing, unpacking, or putting up. It must be possible both to set the magnet bar at liberty, and to secure it again while in its case, as is done in the common compass, and the torsion of the thread must not be altered in so doing; the access of air must be quite cut off even from the mirror, which may be observed through a thin plate of mica, if a piece of plane glass ground parallel is not to be obtained. It is very advantageous to make the case entirely of copper, and even of strong copper-plates, not only for the sake of the increased solidity given

[279][Note by AKTA:] *Der magnetischen Kraft des Magnetometers*. That is, the magnetic moment of the magnetometer.

[280][Note by ER:] Wilhelm Weber's *Werke*, Vol. II, p. 41.

[281][Note by AKTA:] [Web37b, p. 88 of the *Resultate* and p. 41 of Weber's *Werke*] with English translation in [Web41a, p. 86] and French translation in [Web38b].

to the whole apparatus, but also because the case will thus act on the inclosed magnet as a damper, and all the measurements may be made with much greater rapidity. The instrument must be so strong and solid, even when used in the open air, that it may carry two arms, which serve for placing the deflecting bar at equal measured distances east and west. These arms being correctly placed, all the preparations for the experiments of deflection which would otherwise be necessary, — namely, placing the measuring bars horizontally, and in a direction perpendicular to the magnetic meridian, and finding the corresponding points on either side of the magnetometer, — are spared, and the experiments are rendered much easier, and require less time.

13.2 Description of the Several Parts

Figure 1 represents the vertical section of the magnetometer in the direction of the magnetic meridian.

The magnetic bar which forms the needle is bored throughout its length, and the opening which is turned towards the telescope is provided with a lens, in the focus of which at the other end there is a cross of wires. This cross of wires is seen in the telescope, when (as is required for determining the true azimuth in the measurement of the absolute declination) it is adjusted to distant objects, and then directed to the lens. This arrangement was proposed by Airy,[282] to make it possible to dispense with the mirror, and to be able to make, with the same telescope, and without displacing the eye-glass, the astronomical, geodesical, and magnetical observations required in measuring the *absolute* declination. In making this measurement the needle must be *reversed*; but in the reversal the optical axis must not alter its relative position in respect to the needle; this is effected in the closed case by means of a key, turned on the outside, and causing the needle inside to perform half a revolution round its axis of length. But this arrangement is inapplicable to observations which require *great changes* in the position of the needle, as in the experiments of vibration and deflection in the measurement of the absolute intensity. It therefore appeared advantageous to employ also a mirror, placed in the same manner as in the bifilar magnetometer, close to the axis of rotation of the needle, and above the copper case, and available however great the deflections may be.

The copper case is seen to have three openings: the first is into a space containing the mirror, and closed towards the theodolite by a plate of glass, through which the light can pass, in the direction shown in the plate from the scale, to the mirror, and thence back to the telescope of the theodolite. The other two openings are nearly at the same height as the magnetic needle and the telescope of the theodolite. The light entering through one of these apertures illuminates the cross of wires which is stretched across the hindmost end of the hollow needle, passes on to the lens at the other end, and thence, parallel to the horizontal direction marked in the Figure, to the telescope of the theodolite with which the cross of wires is observed. The needle, bored throughout its length, is made accurately cylindrical, and is inclosed in a cylindrical brass box, on the under surface of which are two small projections which fit into two cavities in the copper case when the suspension thread of the needle is let down. The brass box can be fixed in this position by two screws brought through the upper part of the copper case: the box being thus held fast, the needle may *first* be drawn out through the opening in the back of the case, and a brass cylinder of the same

[282][Note by AKTA:] George Biddell Airy (1801-1892).

form as the needle, inclosing a weak magnet, may be placed in its stead, to try the torsion of the thread. *Secondly*, for the purpose of measuring the error of collimation, the needle may be turned in the box round its longitudinal axis, by means of a key introduced through the aperture in the back of the case. During the observations the apertures in the front and the back of the case are closed with a plate of mica to guard against currents of air.

Figure 2 represents a somewhat different and more simple construction of the same instrument; the needle is not hollow, is not enclosed in a brass case, and cannot be reversed. This simplification is admissible when the use of the instrument is to be restricted to the experiments which are to be made *in the open air*, as detailed in the sequel. In this case the mirror is included in the copper case, and its normal forms a right angle with the magnetic axis of the needle. The glazed opening in the side of the case does not impair its action as a damper, and the opening may be made of any convenient size.

Figure 3 represents the outside box, in which the instrument is packed for travelling, and which serves also for suspending the deflecting bar when it is to be used for the experiments of vibration.

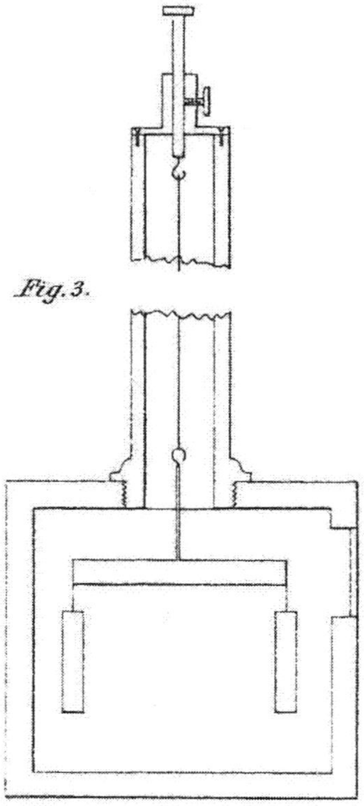

Fig. 3.

A mirror is fixed to the end of the bar, so that it may be observed from a distance with a telescope and scale. The box has a small opening which can be closed with a plate of mica admitting the light. The Figure shows the bar suspended in the box, and loaded with two cylindrical weights, made of brass, and connected by a silk thread passing over a bar parallel to the needle, to keep the centres of gravity of the two weights exactly the length of the bar from each other. The weights serve for the deduction of the moment of inertia.

The unifilar suspension of the bar can be changed for a bifilar, if the variations of the intensity are to be observed. The box must then be placed relatively to the theodolite and to the magnetometer in the manner represented in the ground plan, Figure 4, namely, so that, according to the rule laid down in the *Resultate* for 1837, p. 22,[283,284] the line connecting the middle of the bar with the middle of the magnetometer needle may form with the magnetic meridian an angle of 35°16′. Thus observations of the variations of the declination and of the intensity may be conveniently combined in this manner by travelling observers.

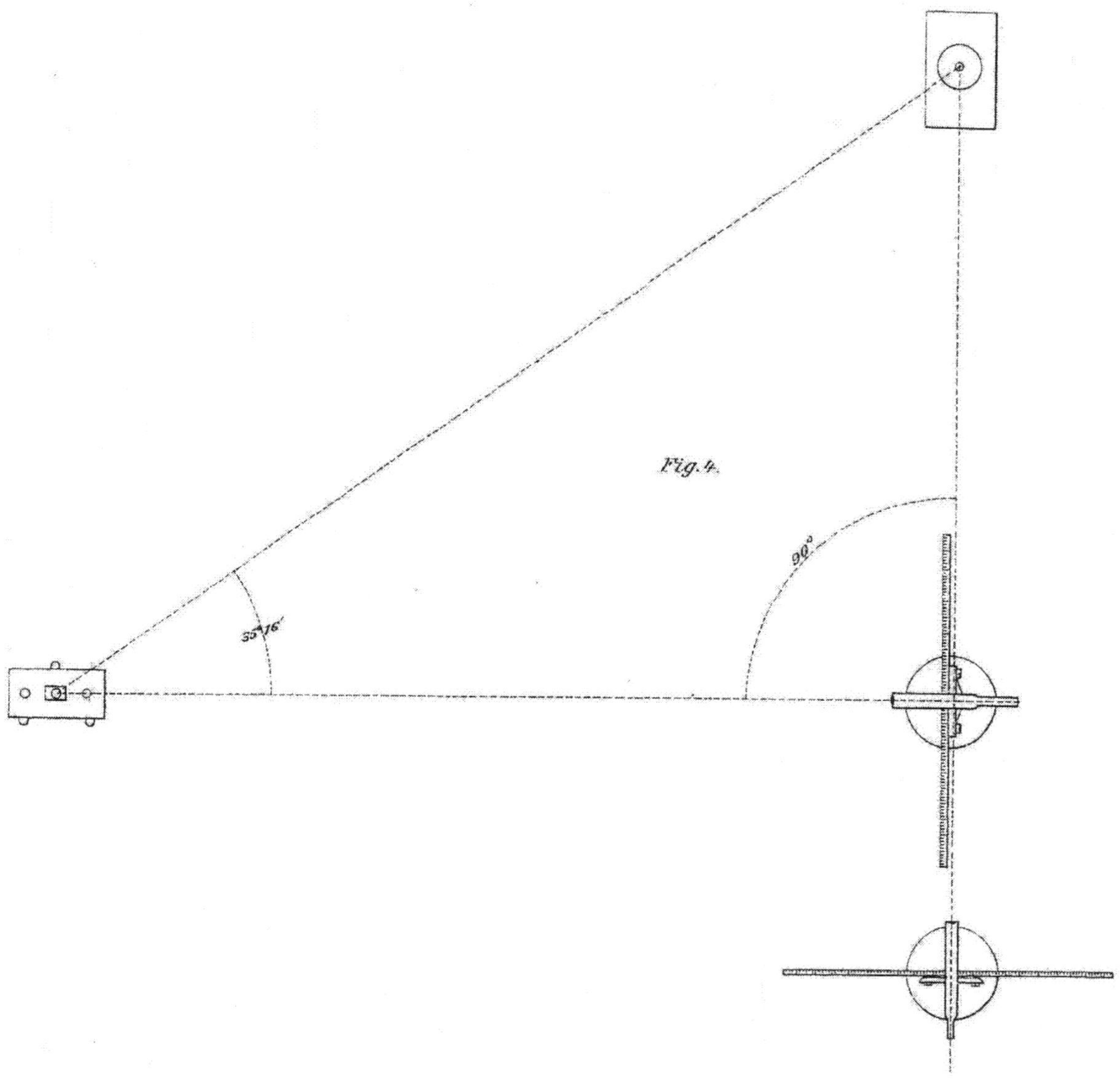

13.3 Examples of Observations and Measurements

13.3.1 Measurement of the Absolute Declination

This measurement resolves itself into three parts: 1. The determination of torsion. 2. The azimuthal determination of the magnetic axis. 3. The azimuthal determination of the true meridian. By the *azimuth of a direction* is here understood the angle formed by two vertical planes, one in the direction in question, and the other in the direction of the optical axis of

283[Note by ER:] Wilhelm Weber's *Werke*, Vol. II, p. 45.

284[Note by AKTA:] [Web38a, p. 22 of the *Resultate* and p. 45 of Weber's *Werke*] with English translation in [Web41d, p. 270] and [Web66b].

the telescope of the theodolite, the alidade being placed on the zero point of the circle.

1. Determination of Torsion

This determination consists of the measurement of the *force of torsion*, and of the *angle of torsion*.

Force of Torsion

There belong to the magnetometer two needles, the magnetic and the torsion needle, which may be suspended to the same thread, and which differ in the proportion of their magnetic moments (M, m). Designating by T the horizontal part of the earth's magnetic force, the force of torsion is to be compared with the force MT, as well as with the force mT.

Comparison with the Force MT

In order to reduce the observations to the same time, the declination was observed in the magnetic observatory simultaneously with both the observations.

Reading of the Torsion Circle	Observation of the position of the Magnetometer by the Scale	Observation in the Magnetic Observatory	Radius in parts of the Scale	Reduced Observation
$355°6'$	275.67	$18°29'49''$	2174	275.67
$175°6'$	237.06	$18°30'42''$		237.31

Hence the force of torsion is given in parts of MT

$$= \frac{57.295}{180°} \cdot \frac{38.36}{2174} = \frac{1}{178} .$$

Comparison with the force mT

Reading of the Torsion Circle	Observation of the position of the Magnetometer by the Scale	Differences	Mean	Radius in divisions of the Scale
269°15′	270.77			
		160.98		
329°54′	109.79			
		171.12		
269°15′	280.91		167.69	2243.5
		168.73		
329°54′	112.18			
		169.94		
269°15′	282.12			

Hence the force of torsion is given in parts of mT

$$= \frac{57.295}{60.65} \cdot \frac{167.69}{2243} = \frac{12.563}{178} .$$

Angle of Torsion

	Observation of the position of the Magnetometer by the Scale	Radius in divisions of the Scale
Magnetic needle	292.90	2174
Torsion needle	328.67	

The distances of the observed divisions of the scale from the zero point of torsion being designated by x and y, then x is the angle of torsion sought, expressed in divisions of the scale; and for determining x we have the following equations:

$$\begin{aligned} 292.90 - x &= 328.67 - y \\ 12.563x &= y . \end{aligned}$$

Hence the angle of torsion in divisions of the scale is found,

$$x = 3.09 ,$$

in seconds of arc

$$x = \frac{3.09}{2174} \cdot 206265'' = 293'' .$$

From this determination of the force and of the angle of torsion, the correction on account of torsion to be applied in measuring the declination is found

$$= \frac{1}{178} \cdot 293'' = 1.65'' .$$

This correction is so small that it may be wholly neglected; the more so, as, during the time occupied in the measurement, the declination itself altered two divisions of the scale, so that the angle of torsion for the time of this measurement almost wholly disappeared.

2. Azimuthal Determination of the Magnetic Axis

In order to reduce the observations to the same time, the declination was observed simultaneously in the magnetic observatory.

	Time: 1839, April 11	Azimuth of the Collimation Line	Observation in the Magnetic Observatory	Reduced Azimuth	Azimuth of the Magnetic Axis
Before reversal	11h 0m	131°22′43″	18°26′26″	131°20′0″	131°41′29.5″
After reversal	11h 37.5m	132°2′59″	18°29′9″	132°2′59″	

3. Azimuthal Determination of the True North

Three visible objects were observed, the positions of which, in respect to the Göttingen Observatory, are given by geodesical measurements.

Designation of the Objects	Distance from the Observatory South	West	Observed Azimuth	Azimuth of the true North
Hohehagen	+6060.00	+12447.70	33°58′50″	150°6′14″
Gartenhaus	+289.28	−27.54	315°17′5″	
Jacobithurm	−710.70	+500.49	117°15′15″	

As there is no correction to be applied on account of torsion, we obtain immediately from hence the westerly declination, by deducting the azimuth of the magnetic axis from the azimuth of the true north.

$$150°6'14'' - 131°41'29.5'' = 18°24'44.5'' ,$$

This result corresponds to 11h 37.5m, 11th April 1839. The declination observed at the same time in the magnetic observatory was

$$18°29'9'' ,$$

showing a difference of $-4'24.5''$, which probably is only in part due to error of observation, and is in part caused by the influence of the copper case surrounding the magnetometer, which may not be wholly free from iron. Repeated measurements, and comparisons with the observations in the magnetic observatory, may serve to deduce such an influence if it exists, so that it may be taken into account in future measurements. A second measurement actually gave a similar result, namely,

1839, April 13	In the open air	In the magnetic observatory
10h 31′	18°18′0″	18°23′36″

showing a difference of $-5'36''$. The mean influence of the copper case in this instrument may therefore be taken as $= -5'$.

13.3.2 Observation of the Variations of Declination

On the 15th April 1839, from 5h 25m to 7h 27.5m, the variations of declination were observed alternately, with the magnetometer in the Göttingen Observatory, and with the small magnetometer. In the following Table the four first columns show the immediate results of observation with the two apparatus. In the final column the observations with the small magnetometer are reduced according to the proportion of the value of the scale divisions. The two series of observations are exhibited graphically in Figure 5, for the purpose of comparison. It may be seen from this example that the observations of the variations of declination can be made with a portable magnetometer with much accuracy.

1839 April 13	Magnetic Observatory A	1839 April 13	Transportable Magnetometer Reading x	Transportable Magnetometer Reduced Value $B = 895 + 3.25(x - 244.2)$
5h 25m	896.00	5h 27.5m	244.95	897.44
5h 30m	895.56	5h 32.5m	244.20	895.00
5h 35m	894.66	5h 37.5m	244.97	897.50
5h 40m	896.47	5h 42.5m	245.20	898.25
5h 45m	899.56	5h 47.5m	246.18	901.44
5h 50m	899.52	5h 52.5m	245.78	900.14
5h 55m	898.78	5h 57.5m	246.02	900.91
6h 0m	900.57	6h 2.5m	247.35	905.24
6h 5m	905.95	6h 7.5m	248.04	907.48
6h 10m	908.00	6h 12.5m	249.77	913.10
6h 15m	916.77	6h 17.5m	251.77	919.60
6h 20m	920.00	6h 22.5m	251.77	919.60
6h 25m	919.66	6h 27.5m	251.56	918.92
6h 30m	916.63	6h 32.5m	250.70	916.12
6h 35m	912.72	6h 37.5m	250.96	916.97
6h 40m	917.66	6h 42.5m	251.74	919.51
6h 45m	927.35	6h 47.5m	254.32	927.89
7h 0m	941.27	7h 2.5m	260.79	948.92
7h 5m	959.33	7h 7.5m	265.71	964.91
7h 10m	964.53	7h 12.5m	261.27	950.48
7h 15m	936.38	7h 17.5m	254.34	927.95
7h 20m	922.80	7h 22.5m	251.75	919.54
7h 25m	914.42	7h 27.5m	250.09	914.14

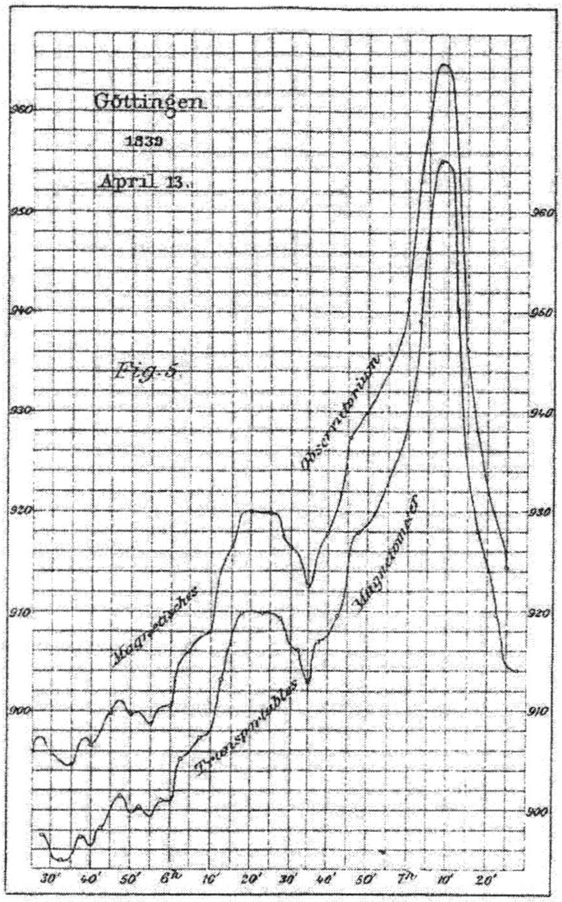

13.3.3 Absolute Measure of the Intensity

The measurement of the intensity divides itself into four parts.[285] 1. The determination of torsion. 2. Of the moment of inertia of the deflecting bar. 3. The experiments of deflection. 4. The experiments of vibration. I will confine myself in this place, for the sake of brevity, to two parts, viz. the determination of the moment of inertia, and the experiments of deflection, which are especially instructive towards a knowledge of the instrument. The determination of torsion has been already spoken of in the measurement of declination, and the experiments of vibration are so simple and so well known, that it is sufficient to give their results.

1. Determination of the Moment of Inertia

The deflecting bar is suspended to a thread or wire, and is then vibrated: 1) without a weight; 2) with a weight, the moment of inertia of which is known.

[285][Note by AKTA:] Friedrich Kohlrausch discussed an example of the measurement of the horizontal intensity of the Earth's magnetism with Weber's portable magnetometer, [Koh83, pp. 172-173].

Vibrations without a weight			
Number of Vibrations	Time	Arc of Vibration	Reduced time of Vibration
0	7h 20m 51.27s	8°56′	
26	7h 23m 45.49s	8°40′	6.698″
61	7h 27m 39.92s	8°8′	6.695″
115	7h 33m 41.64s	7°22′	6.696″
151	7h 37m 42.80s	6°56′	6.695″
186	7h 41m 37.19s	6°32′	
Vibrations with a weight			
0	2h 18m 35.57s	8°16′	12.058″
46	2h 27m 50.45s	6°58′	12.039″
125	2h 43m 41.76s	5°4′	12.019″
200	2h 58m 43.31s	3°20′	

Hence the mean time of vibration without a weight is = 6.696″, and with a weight 12.039″. For determining the moment of inertia of the weight we have the following data: 1) the length l of the deflecting bar, or the distance apart of the threads which hang from its two ends and support two equal cylindrical weights; 2) the mass $2p$; 3) the radius r of the two cylinders.

$$\begin{aligned} l &= 93.42\ mm\ , \\ 2p &= 50000.00\ mg\ , \\ r &= 4.60\ mm\ . \end{aligned}$$

If the mass of the cylinder were concentrated in its axis, its moment of inertia would be

$$\frac{1}{2}l^2p = 109091000\ .$$

If the cylinders revolved only round their own axis, their moment would be

$$r^2p = 529000\ .$$

Their moment in the above experiments is to be taken as equal to the sum of

$$\frac{1}{2}l^2p + r^2p = 109620000\ .$$

Whence therefore the moment of inertia of the oscillating bar may be obtained from the equation

$$MT = \frac{\pi^2 K}{t^2} = \frac{\pi^2(K+K')}{t'^2}\ ,$$

where K' signifies the known, and K the desired moment of inertia, t' the time of vibration with a weight, and t the time of vibration without a weight, consequently

$$K = 49103000\ .$$

In these experiments the needle was suspended to a thread in which the force of torsion was so small as to be insensible. The same series of experiments was repeated with the needle suspended by a wire in which the force of torsion was much greater; the result was almost the same as before, namely,

$$K = 49044000 \; .$$

Finally, in order to furnish a check, the deflecting bar was weighed, and its length and radius were exactly measured:

$$\begin{aligned} \text{Weight } p' &= 66670 \; mg \\ \text{Length } l &= 93.42 \; mm \\ \text{Radius } r &= 5.45 \; mm \; , \end{aligned}$$

whence its moment of inertia may be calculated. Supposing perfect internal homogeneity,

$$K = \frac{1}{12} l^2 p' + \frac{1}{4} r'^2 p' = 48982000 \; .$$

The accordance of all these experiments sufficiently shows that the moment of inertia of even such small bars may be determined with great precision.

2. Experiments of Deflection

1839, February 13 Distance in Millimeters	North Pole	Readings	Double Deflection In divisions of the scale		Arc values
−556.75	E.	372.95			
			240.62		
	W.	132.33		241.03	5°30.3′
			241.45		
	E.	373.78			
−453.25	E.	475.91			
			447.55		
	W.	28.36		447.89	10°9.3′
			448.22		
	E.	476.58			
+453.25	E.	480.04			
			448.21		
	W.	31.83		448.32	10°11.2′
			448.44		
	E.	480.27			
+556.75	E.	375.93			
			240.87		
	W.	135.06		240.82	5°30.0′
			240.76		
	E.	375.82			

Hence the simple deflections v_0, v_1 are obtained for the distances R_0, R_1 (without regard to signs)

$$v_0 = 2°45'4.5'' \,, \quad \text{for} \quad R_0 = 556.75$$
$$v_1 = 5°5'7.5'' \,, \quad \text{for} \quad R_1 = 453.25 \,.$$

Consequently, if $\tan v$ be developed according to the powers of R,

$$\tan v = 8305800R^{-3} - 4081300000R^{-5}$$

whence (see *Intensitas Vis Magneticae*, Sections 21, 22),[286]

$$\frac{M}{T} = 4152900 \,.$$

With the comparatively great distance of the deflecting bar from the needle (equal to from 5 to 6 times the length of the needle), the determination of the coefficient of the second member of this equation (which is to be divided by the 5th power of the distance) is uncertain, and it is therefore better to disregard it. We then obtain for M/T two values,

[286][Note by AKTA:] See Chapters 5, 6 and 7, especially Sections 7.21 and 7.22.

$$R_0^3 \tan v_0 = 4146600$$
$$R_1^3 \tan v_1 = 4143200 ,$$

viz. the mean of which may be taken, consequently,

$$\frac{M}{T} = 4144900 ,$$

which differs but little from the above value.

If to the results obtained we add lastly the time of vibration t, which was found to be[287]

$$t = 60.586'' ,$$ [288]

and if we assume $K = 49073500$, we obtain

$$MT = \frac{\pi^2 K}{t^2} = 13195000 ,$$

consequently

$$T = 1.7842 .$$

We are not enabled to test and compare this result further, as a simultaneous measurement with the large magnetometer could not be executed at that time. When a new measurement of the earth's magnetic force is made in the Göttingen Observatory, the opportunity of comparison thus afforded will not be neglected.[289]

The improvements, (represented in Figures 2, 6, 7, 8,) which, since the above was written, I have caused to be made in the transportable magnetometer, are designed to facilitate the use of the instrument in the open air, as in travelling it will be rare to meet with a suitable building free from iron for the execution of absolute measurements. It is not absolutely necessary that the whole of the observations for these purposes should be made in the open air; and on account of the liability to interruption from weather, it is desirable to reduce the number requiring this exposure as much as possible. In the improved construction I have given great care and consideration to this part of the subject, and have found it possible to arrange the observations in such manner that the greater part may be made in a room, including those which would be made to the greatest disadvantage in the open air.[290]

[287][Note by AKTA:] In the English translation the next value of the time of vibration t was written as:

$$t = 6'' \cdot 0586 .$$

We wrote its value according to the original German text, namely, $t = 60.586''$.

[288][Note by WW:] The bar having been magnetized afresh for the experiments of vibration and deflection, had a shorter time of vibration than in the previous experiments on the moment of inertia.

[289][Note by AKTA:] This is the end of Weber's 1839 paper, [Web39], with the exception of a last paragraph on the cost of these instruments, which has not been translated. The following text comes from manuscript communications from Wilhelm Weber to Edward Sabine (1788-1883), [Web41e] and [Web66c].

[290][Note by AKTA:] A larger plate appears on page 182.

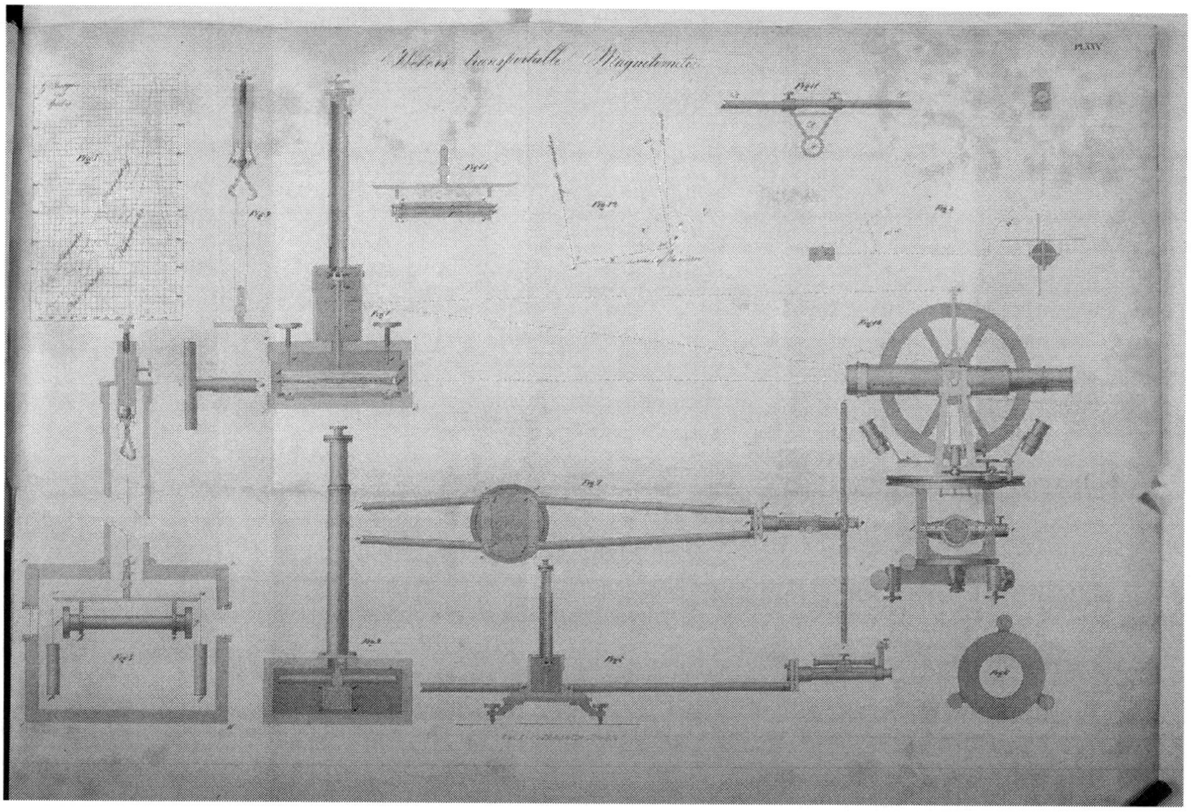

Figure 6 represents the tripod stand, on which the measuring apparatus, Figure 7, and the magnetometer, Figure 2, are to be placed and levelled, as shown in Figure 8.[291]

The measuring apparatus, Figure 7, required for the deflection experiments, consists of a copper-plate fitting on the tripod, and carrying the supporters of the deflecting-bar; each of these is formed of two converging tubes connected at their extremities, from whence proceeds a third tube provided with a graduation, and on this the deflecting-bar is to be placed: this tube forms also the reading telescope, and has the reading scale attached to it.

Figure 8 represents the magnetometer placed on the measuring apparatus, which rests itself upon the tripod: the needle is suspended in a copper case, which acts as a damper in checking the vibrations. The mirror close below the needle is directed to the east. The whole of the eastern side of the copper case can be removed, to give access to the screw to which the suspension is fastened, and by which the inclination of the mirror may be corrected. In the middle of this side is an opening closed by a piece of plane glass, making a small angle with the vertical, in order that the reading telescope, which is directed to the mirror behind the glass, may not see a double image of the scale.

For the *measurement of the absolute intensity* the deflection experiments alone require to be made in the open air; the remainder may be made in a room if more convenient; for if the magnetism of the needle, which can be ascertained in a room, be known, the intensity of the earth's magnetism may be calculated from that of the needle, and from the experiments of deflection made in the open air.[292] It should be noticed, however, that the determination of the magnetism of the needle in such cases requires a complete measurement of the intensity to be gone through, including both the experiments of deflection and those of vibration, with and without the weights. The magnetism of the needle should also be determined

[291][Note by AKTA:] These Figures appear on page 173 and the following.

[292][Note by WW:] The experiments of vibration might be made in the open air instead of those of deflection; but in such case the instrument would afford less certainty and less convenience.

either shortly before, or shortly after, the deflection experiments in the open air, because it is liable to alteration: and the temperature in the room and in the air should be as nearly the same as possible.

The experiments of deflection in the open air require only a solid foundation, on which the tripod may be placed and levelled; the measuring apparatus, resting on it, carries the deflecting bar, the telescope, and the scale, each in its due position relatively to the others; and the whole system can be turned upon the tripod without their displacement. The copper case of the magnetometer fits into the depression *ab*, Figure 8, by which its position is fixed relatively to all the other parts. The whole instrument is then turned on the tripod until the middle of the scale is seen in the reading telescope, and it is then ready for the deflection experiments.

The vernier of the deflecting bar being placed on the zero point of the graduation of the measuring apparatus, the deflection of the needle is observed. The deflecting bar is then reversed, and the observation repeated. The bar is then removed to the end of the measuring apparatus, and the vernier set to 1000 mm of the graduated scale, when the deflected position of the needle is again observed before and after the reversal of the bar. Let the four observed deflections be called m, m', n, n', — the absolute intensity of the magnetism of the needle, previously observed in a room, M, — and the arc-value of a division of the scale, determined also in a room (the torsion being taken into account), α, — then the absolute horizontal intensity of the earth's magnetism will be

$$I = \frac{2}{500^3} \cdot \frac{M}{\tan v} \ ,$$

where

$$v = \frac{1}{2} \arctan \frac{1}{2} (m - m' + n - n')\alpha \ .$$

This simple formula may be employed, because the small dimensions of the needle and bar, relatively to their distance apart, renders the next member (having the fifth power of the distance in the denominator) insensible.

Figure 10 represents the theodolite used in observing the declination and its variations;[293] it is provided with a verification telescope, having a small scale at the end: a larger scale is placed above the theodolite, perpendicular to the optical axis of the principal telescope.

The observation of the absolute declination may be divided into those parts which must be made in the open air, and those which may be made in a room. Figure 11 represents in A the cross-section of the tube of the magnetometer telescope, and in BC the scale; between A and BC is a transparent space; the theodolite must be so placed that the observer may look with the verification telescope through the space D towards the mirror of the magnetometer needle, and perceive the image of the scale attached to that telescope; he must first observe the position of the needle by this scale, and thence determine the angle ϕ (Figure 12), which the optical axis of the verification telescope makes with the normal to the mirror of the magnetometer; he must then bisect objects of known azimuth with the principal telescope of the theodolite, and thence find the angle ψ, corresponding on the divided limb to the direction of the principal telescope relatively to the north.[294]

[293] [Note by AKTA:] This Figure 10 appears on page 179.

[294] [Note by AKTA:] These Figures appear on page 179 and the following.

These are all the observations required to be made in the open air in determining the declination. The angle χ, Figure 12, corresponding, on the graduated limb, to the parallel position of the optical axes of the two telescopes of the theodolite, can be ascertained in a room; as can also the angle ρ which the magnetic meridian makes with the normal of the mirror belonging to the needle. Hence we obtain

$(\chi - \psi)$ the angle which the optical axis of the verification telescope makes with the true meridian.

$(\chi-\psi)-\phi$, the angle which the mirror-normal of the needle makes with the true meridian.

$\rho - \{(\chi - \psi) - \phi\}$ the angle which the magnetic meridian makes with the true meridian.

The angle χ is found by placing a plane mirror before the verification telescope, and viewing in the telescope the reflected image of a vertical thread suspended over the middle of the object glass; a vertical thread is also suspended over the middle of the principal telescope, and the telescope adjusted to its reflected image; the reading on the circle gives the angle χ, supposing the collimation error of the principal telescope to remain unaltered when the eye-piece is adjusted to distant objects; otherwise the alteration must be sought by reversing the telescope, and applied as a correction to the reading on the circle.

The angle ρ is determined by directing the principal telescope of the theodolite from B (Figure 12) to C, a second needle suspended in the wooden case, as represented in Figure 3; the verification telescope is directed on the first needle A, in the copper case as in the open air.

The needle C is furnished either with a collimator or a mirror, and is capable of reversal. The direction of its magnetic axis is next to be found, i.e. the angle μ, to which the theodolite must be adjusted, in order that the optical axis of its principal telescope may be parallel with the direction of the magnetic axis of the needle C, whence the angle ρ $(= \pi - (\chi - \phi) + \mu)$ is obtained, if the two needles A and C are sufficiently distant apart to exert no sensible influence on each other, so that their magnetic axes may be regarded as parallel. But if this be not the case, it is easy to determine the angle ν formed by the magnetic axes of the two needles,[295,296] and to add it as a correction to the value, as above, of ρ; i.e.

$$\rho = \pi - (\chi - \phi) + \mu + \nu \ .$$

The suspension of the needle in the wooden case is so contrived, that it may be used either as an unifilar or as a bifilar magnetometer. This contrivance is represented in Figure 9.[297] The variations of the declination and of the horizontal intensity can thus be observed at the same time; the former with the magnetometer in the copper case, and the latter

[295][Note by WW:] From the propositions contained in the *Resultate* for 1837, page 22 *et seq.*, it follows that if $ABC = 90°$, $ACB = \alpha$, and $AC = \nu$, and if m and m' denote respectively the magnetism of the needles A and C,

$$\nu = \frac{3}{2} \sin 2\alpha \cdot \frac{m - m'}{r^3 T} \ .$$

The value of m and m' must be determined by the deflections δ and δ' of a compass needle placed successively east and west at the distance d, namely

$$\frac{m}{T} = \frac{\delta d^3}{2} \ , \qquad \frac{m'}{T} = \frac{\delta' d^3}{2} \ .$$

[296][Note by AKTA:] [Web38a, p. 22 of the *Resultate* and p. 45 of Weber's *Werke*] with English translation in [Web41d, p. 270] and [Web66b].

[297][Note by AKTA:] This Figure 9 appears on page 178.

with the magnetometer in the wooden case. In preparing for the latter observations, the telescope of the theodolite is to be directed perpendicularly to the magnetic meridian, and the magnetometer in the wooden case is to be placed in the same direction. The time of vibration t of the needle, with the unifilar suspension, must be determined, if not already known, which it will generally be, from the experiments of vibration belonging to the measurement of the absolute intensity. The unifilar suspension must then be changed for the bifilar without altering the direction of the magnetic axis, and the time of vibration must be observed afresh, the distance apart of the suspension threads being increased until t' is about $= 0.6871t$. The torsion circle must then be turned until the middle of the scale appears in the field of view of the telescope, and the time of vibration t'' observed. The magnetometer is then in the transversal position proper for observing the variations of intensity, and the value of the scale divisions may easily be calculated from the observed times of vibration t, t', t''; namely, if σ denote the arc-value of a division of the scale in parts of radius, the value of a division of the scale, in parts of the whole horizontal intensity, is

$$\frac{t^2}{t''^2} \cdot \sigma = \sqrt{\left(1 - 2\frac{t'^2}{t^2}\right)} \cdot \frac{t^2}{t'^2} \cdot \sigma \, .$$

Explanation of the Figures

Figure 1 a, b, c, d is a vertical longitudinal section of the copper case of the magnetometer, with the needle e, f suspended by a silk thread g, h.

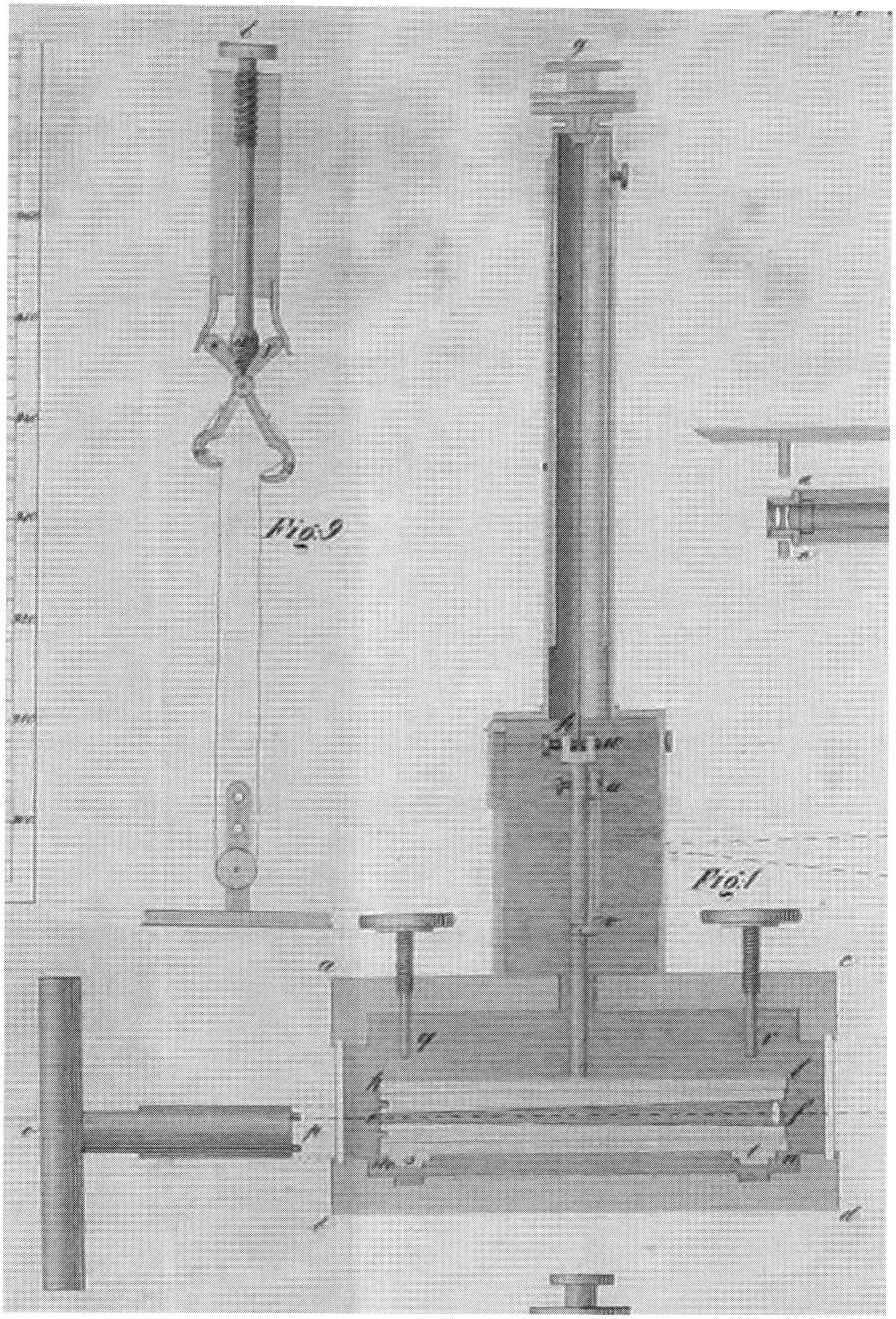

The needle is seen to be pierced through its length, and provided at the extremity f with a lens; it is inclosed in a copper tube k, l, m, n, and can be turned by means of a key o, p, which is accessible by an opening in the copper case. In doing this the copper tube is held by two screws q, r, and two projections s, t. The mirror u, v is seen above the copper case, near the axis of rotation of the needle. A dotted line indicates how the telescope of the theodolite, Figure 10, is directed, both to the needle and to the lens at its end f, and also to the mirror u, v.

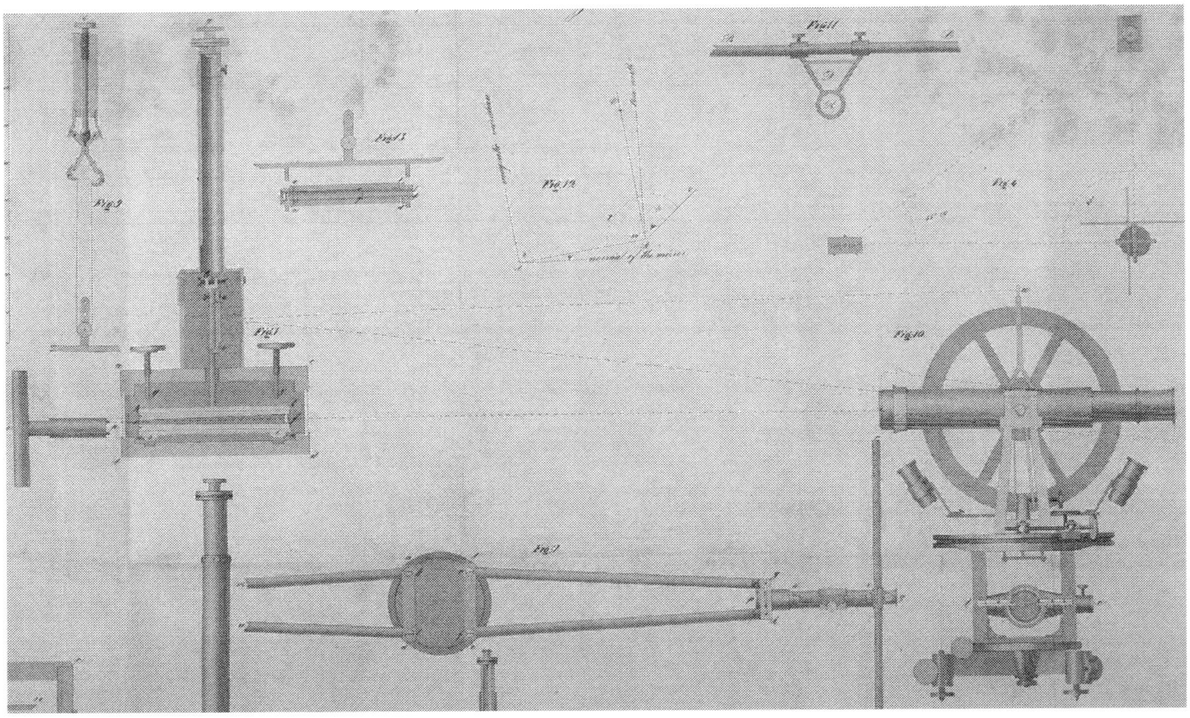

It is also seen how the inclination of the mirror may be regulated by the screw w, that the image of the scale placed above the telescope at a, Figure 10, may appear in the field of view. This Figure is half the size of the instrument itself.

Figure 2 represents a magnetometer, which differs from the one just described in not being adapted for *complete* measurements of the declination. The collimator is omitted, and the needle cannot be reversed. The mirror a, b, c, d is inclosed in the copper case, and is parallel to the plane of the magnetic meridian; the inclination of the mirror is regulated by the screw at e; the copper case forms an unbroken damper round the needle, except at the aperture for the suspension thread; the mirror is observed through a glass plate in one of the sides of the copper case. This Figure is also half size.

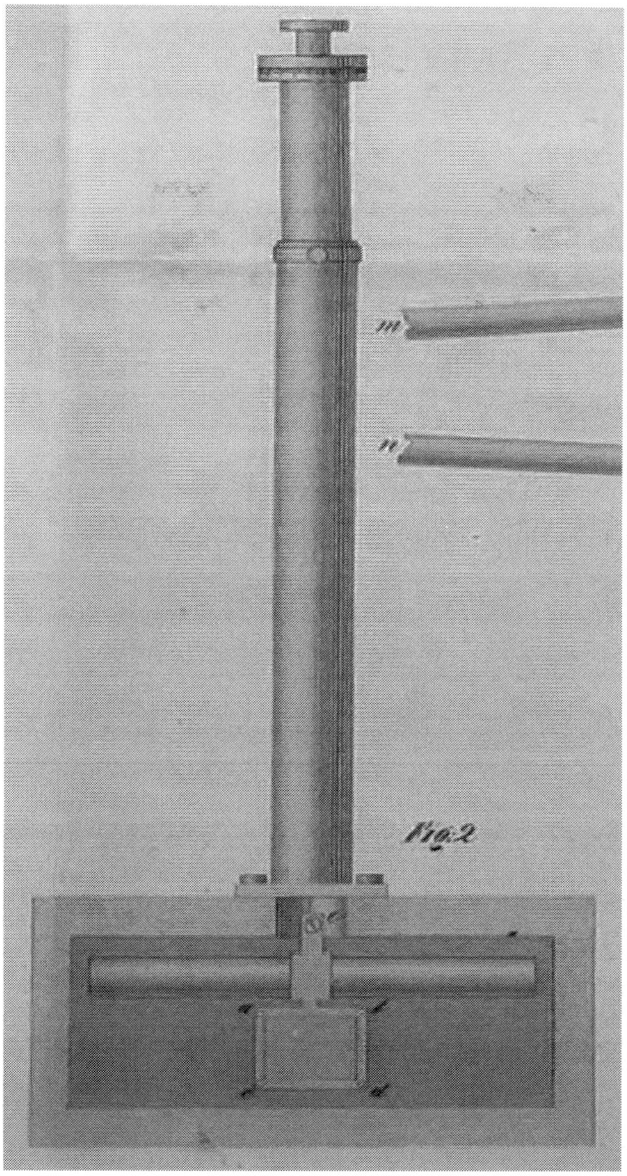

Figure 3 a, b, c, d, e represents the wooden case, in which either of the instruments shown in Figures 1 and 2 are packed for travelling.

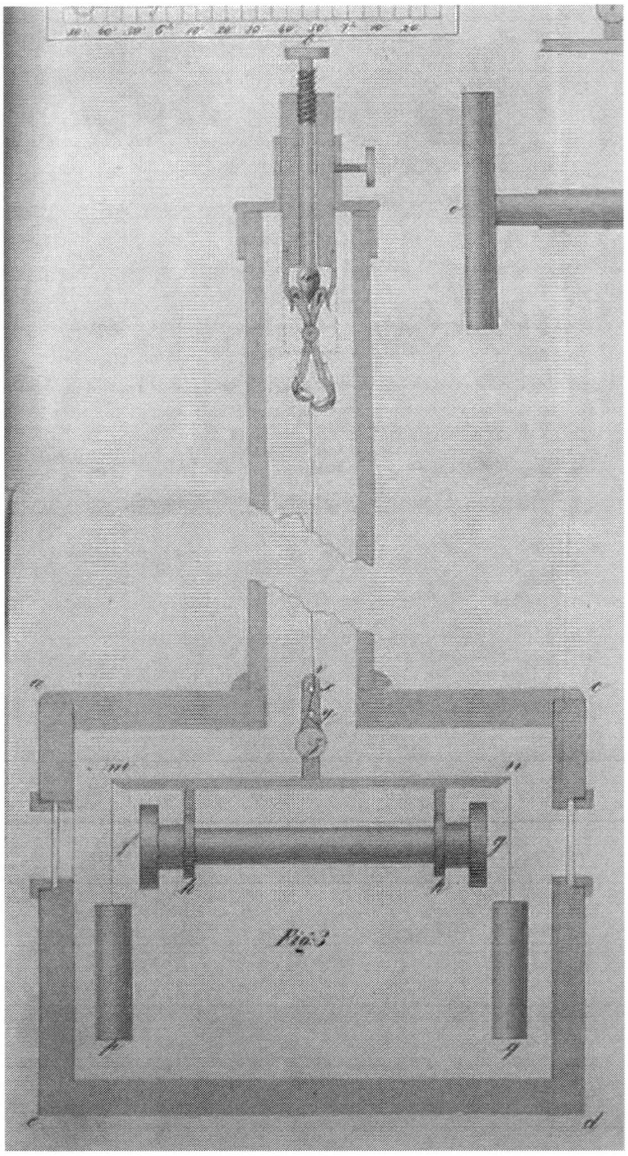

The lid, with the tube a, b, c which is fastened to it, is taken off, the instrument placed inside, and the box closed again. When observations are made, this box serves for suspending a second needle, the time of vibration of which is required for the measurement of the absolute intensity; this second needle f, g is provided at both ends with mirrors, one of which serves for observing the scale. The needle rests on two supports h, k, attached to a small measuring bar m, n, over which passes a thread carrying the weights p, q, which serve to increase the moment of inertia of the oscillating needle. The needle can be turned in the supports h, k, and may be reversed; rendering it available, in absolute measurements of the declination, as an auxiliary needle, when the instrument represented in Figure 2 is used, the needle of which is not reversible. For this purpose, instead of a needle with a mirror, one with a collimator, Figure 13, may be placed in h, k. It consists of a magnetic steel tube a, b, c, d, carrying at the end a, c, an achromatic object-glass; and at its other extremity a sliding tube of brass e, f, g, h, provided with a glass micrometer in the focus of the object-glass. It will be seen also by Figure 3 that this needle is suspended to two threads, the upper points of attachment of which are r and s. The threads are conducted over a roller z to give them equal tension, and are united in one from u to v, forming an unifilar suspension, which may be converted into a bifilar by opening out the apparatus α, β, γ, δ, which is done by pressing down the knob

w by the screw t, and disengaging the threads from the pins x, y, as represented in Figure 9. Figure 3 is also half size.

Figure 4 A, is the theodolite carrying two telescopes and two scales; one telescope and one scale serve for observing the unifilar magnetometer B, and the other telescope and scale for observing the bifilar magnetometer C. The Figure gives the angles which the instruments ought to form with each other.

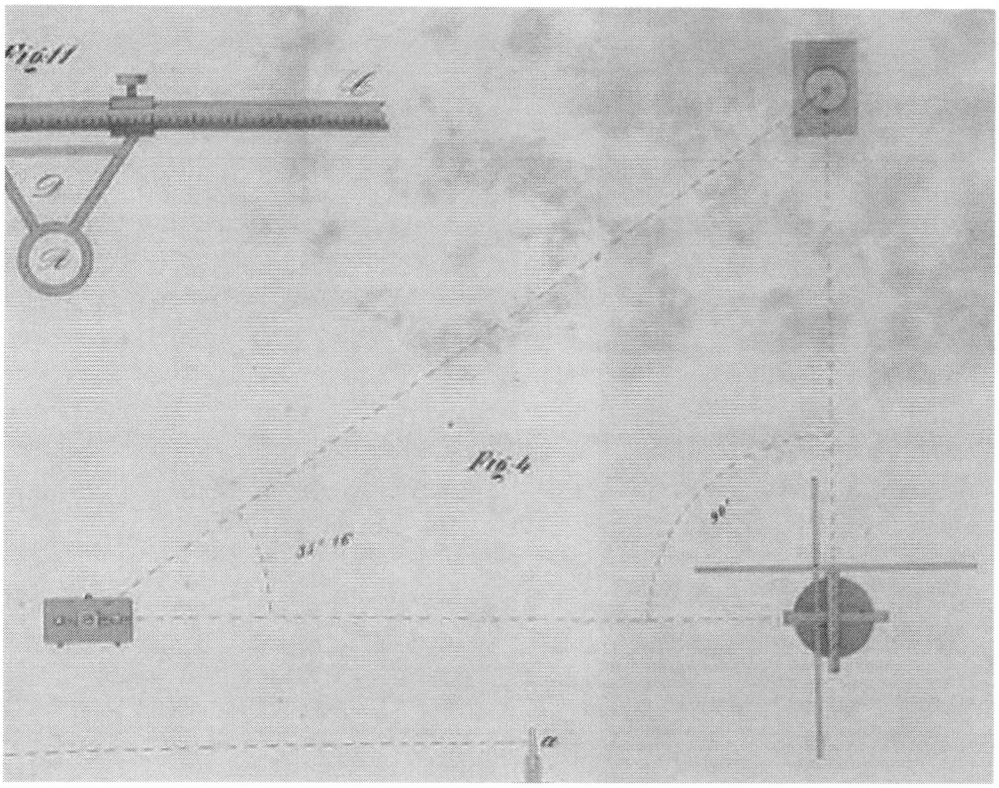

Figure 5 is a graphical representation of the variations of the declination observed on the 13th of April, 1839, at Göttingen, simultaneously in the magnetic observatory, and with the transportable magnetometer.

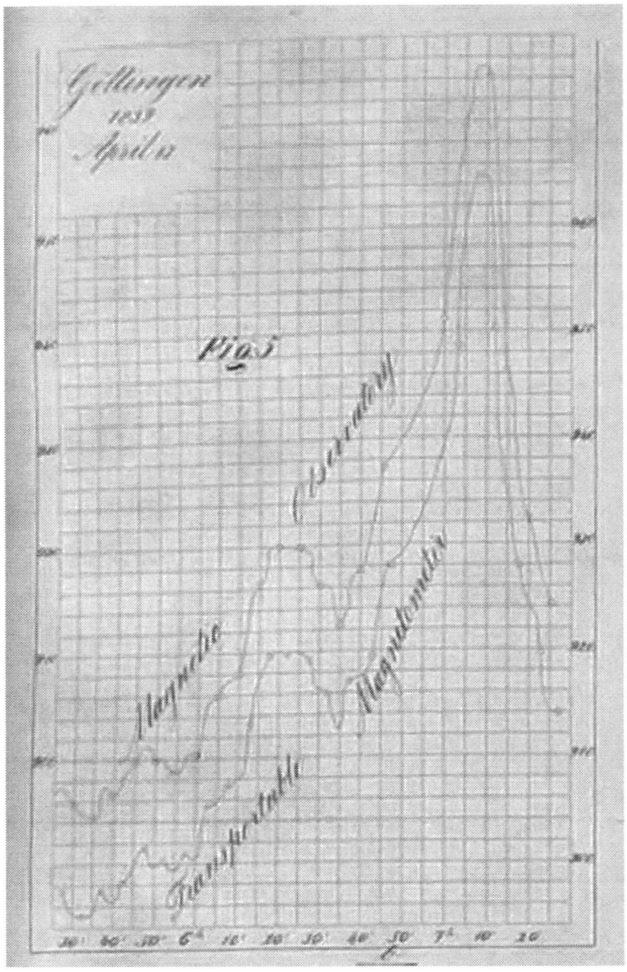

Figure 6 is the tripod on which the magnetometer, Figure 2, is placed and levelled.

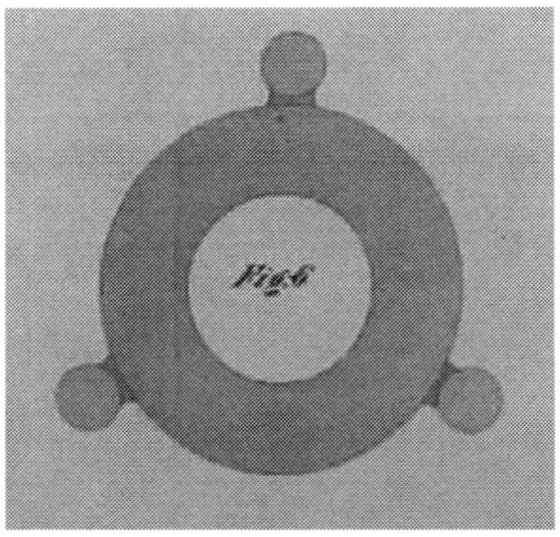

Figure 7 is the apparatus required for the experiments of deflection. a, b, c, d is a copper disc which fits on to the tripod, Figure 6; e, f, g, h, and k, l, m, n, are arms screwed to the copper disc at e, f and k, l; one arm carries the telescope p, q, to which the scale r, s is attached, and upon which the deflecting bar u, v is to be laid; the other arm carries a tube on which the deflecting bar is also laid, but which could not be conveniently represented in the Figure. Between e, f and k, l the magnetometer (Figure 2) is placed.

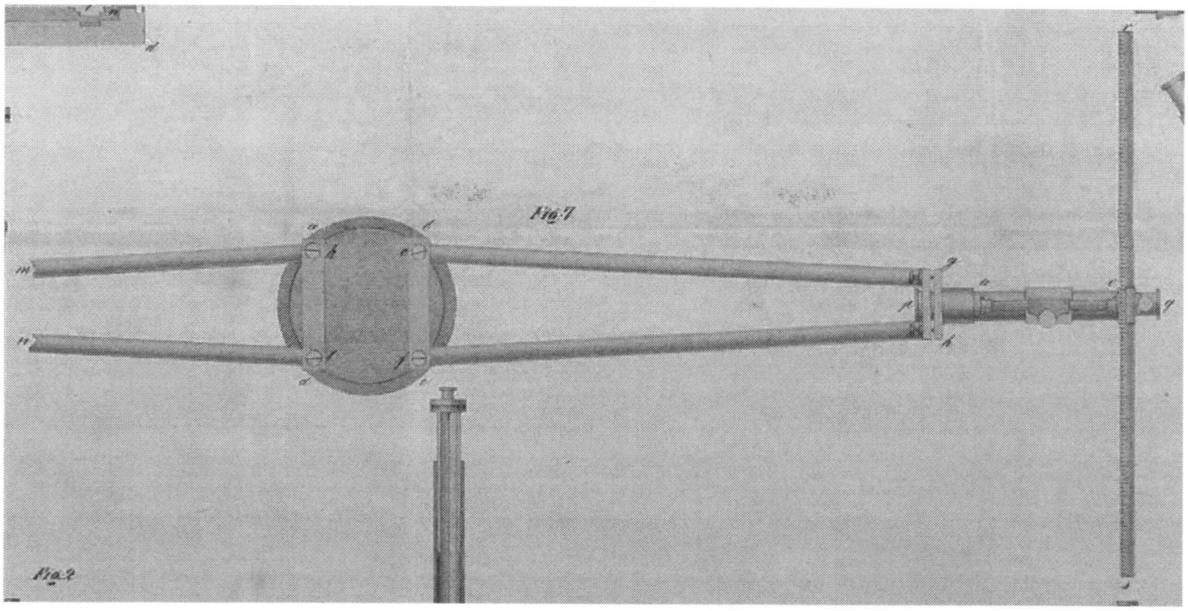

Figure 8 is a smaller side-view of the magnetometer represented in Figure 2, in its proper relative position to the measuring apparatus, Figure 7, and resting on the tripod, Figure 6. In this view the needle is seen only by its circular cross-section, and the glass plate is shown, in the side of the case which permits the image of the scale, reflected from the mirror, to be observed with the telescope.

Figure 9 is explained in the description of Figure 3.

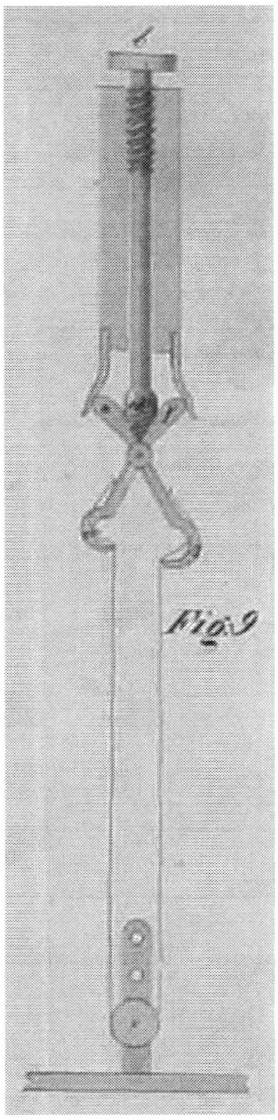

Figure 10 represents the theodolite with the verification telescope: two scales are seen, one of which, *a*, is applied in such manner that its middle corresponds to the prolongation of the vertical axis of rotation of the theodolite; the other, *b*, *c*, is attached in front of the object-glass of the verification telescope. It is very narrow, in order to intercept the less light.

Figures 11, 12 and 13 are explained in the text.

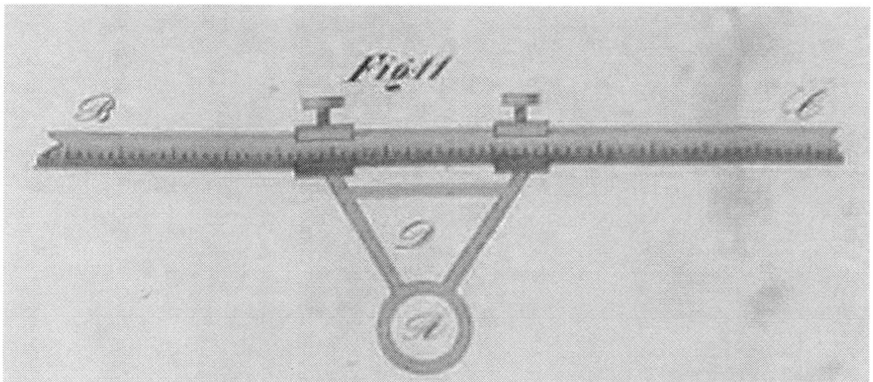

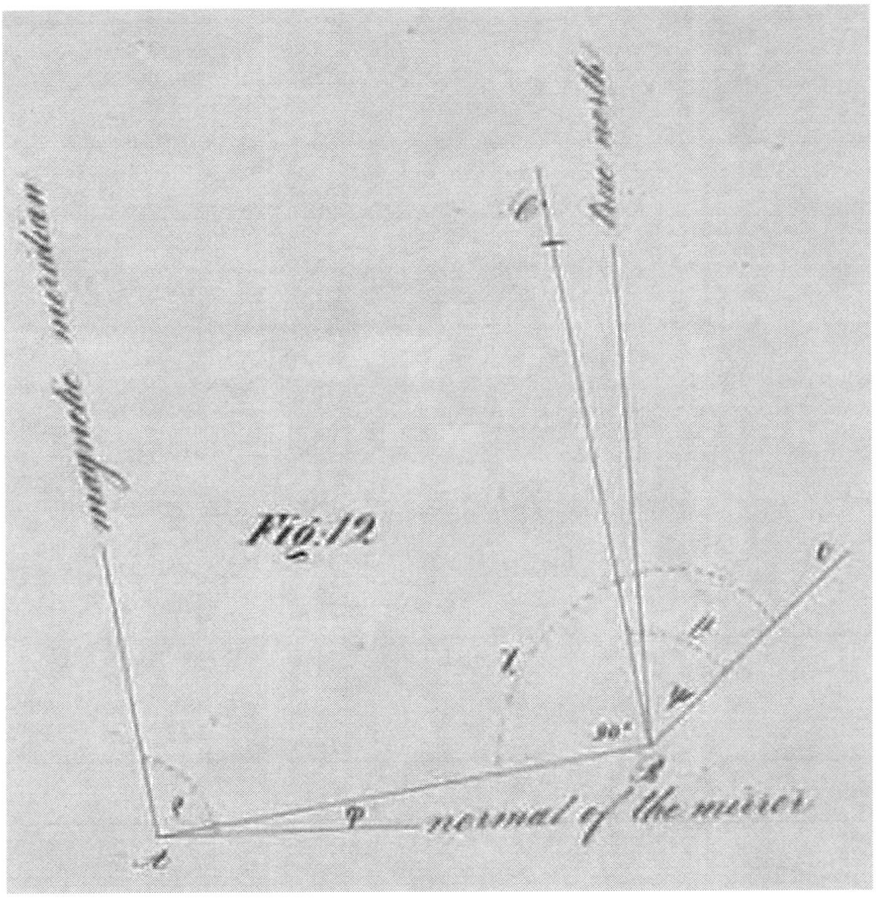
Fig. 12
magnetic meridian
true north
90°
normal of the mirror

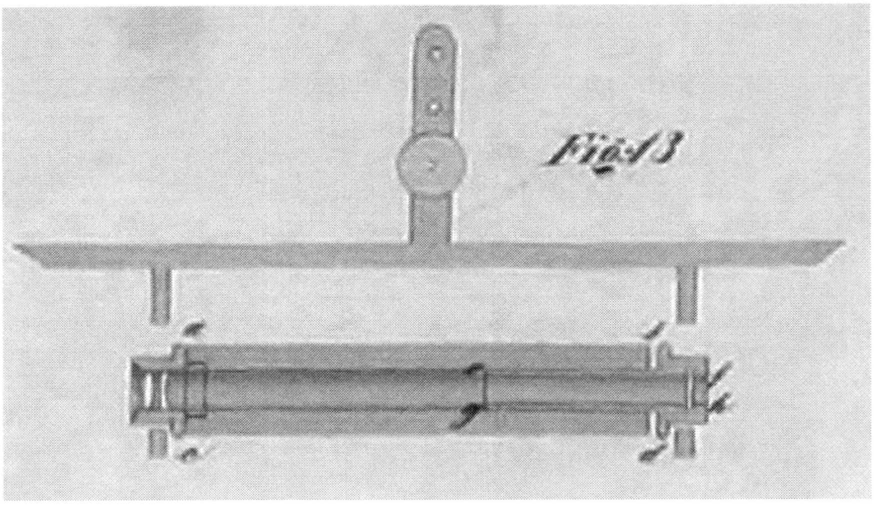
Fig. 13

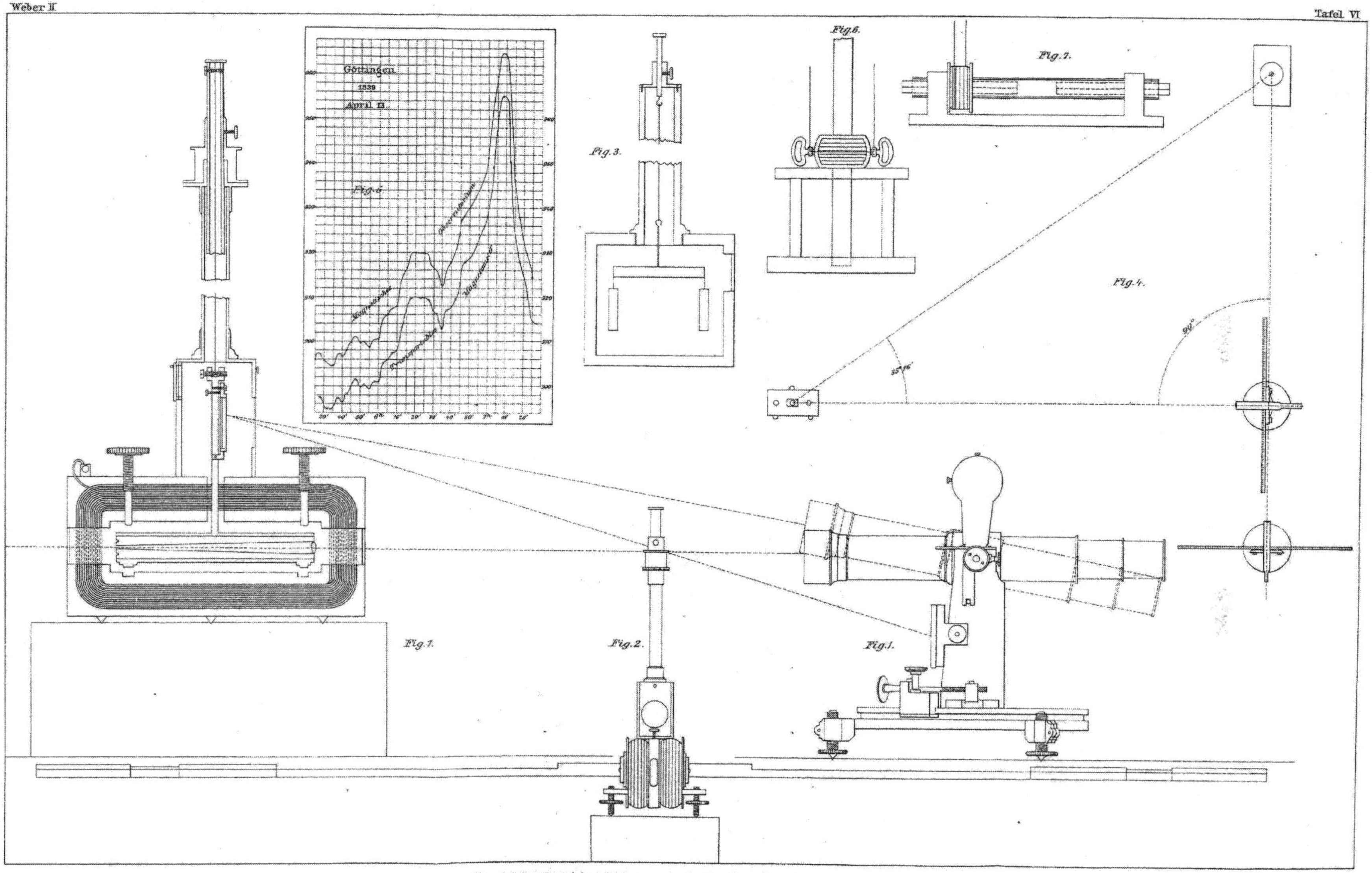

Maasstab: 0,85 der Original-Zeichnungen in den „Resultaten“ vom Jahre 1839.

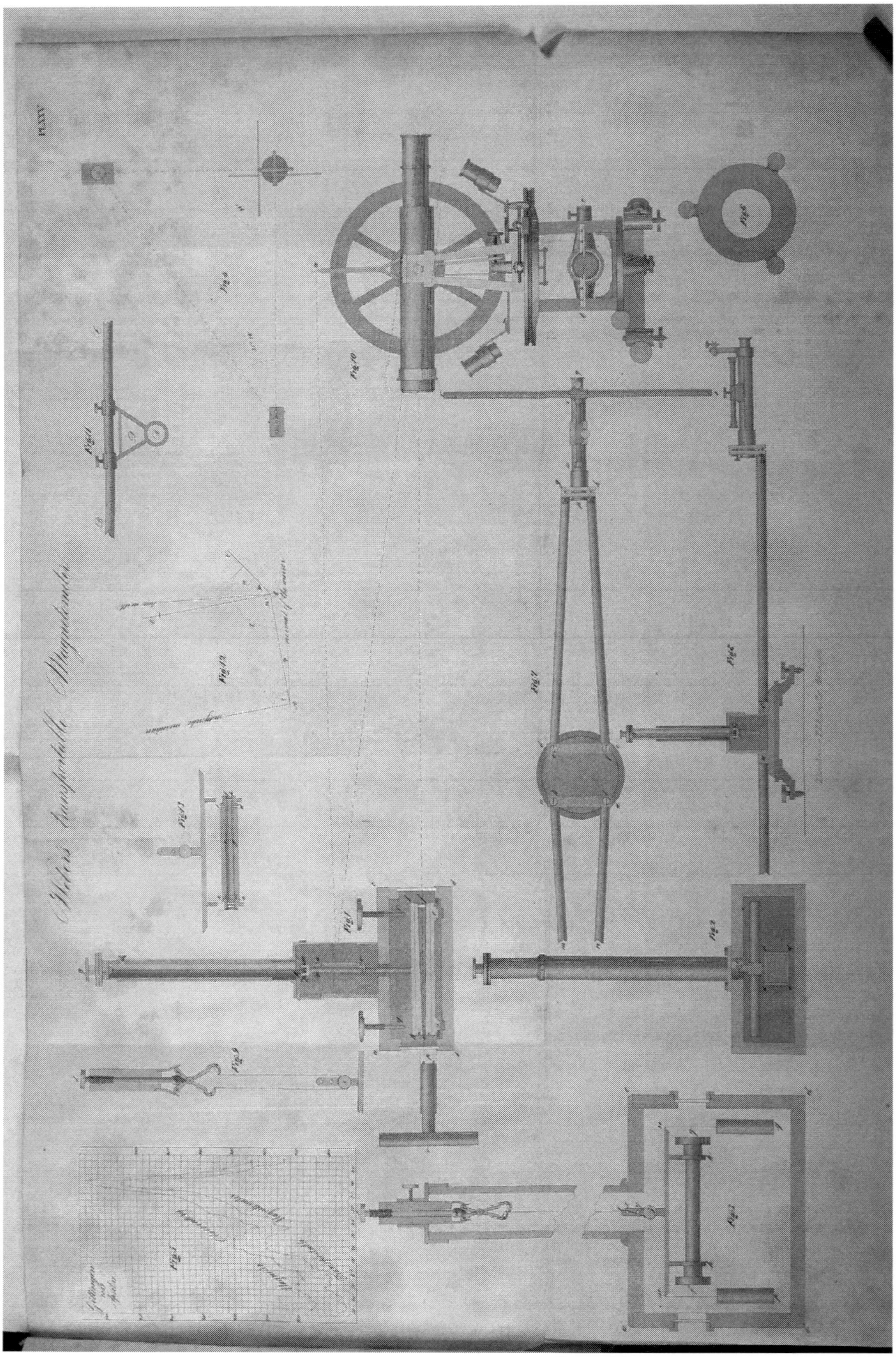

Pl. XXV
Webers transportable Magnetometer

Chapter 14

[Wöhler and Weber, 1841] Composition of Galvanic Piles

[Friedrich Wöhler and Wilhelm Weber][298,299,300,301]

Professors Wöhler[302] and Weber informed the *Königl. Gesellschaft der Wissenschaften* (*Royal Society of Sciences*), together with some remarks, about a discovery made by Professor Poggendorff[303] on the composition of galvanic piles and submitted them to the *Königl. Akademie der Wissenschaften zu Berlin* (*Royal Academy of Sciences of Berlin*) on April 29th of this year.[304]

It is known that in order to produce the greatest galvanic effects, it is no longer necessary to use giant devices that are as uncomfortable as they used to be, but that in recent times one has learned to achieve the same effects with small and convenient devices. The best performance is shown by a pile, described by Mr. Grove,[305] where small clay cells, the walls of which are permeated with liquid, are filled with ordinary nitric acid and externally surrounded with dilute sulphuric acid. In the former liquid platinum plates are immersed, amalgamated zinc plates are immersed in the latter, and the necessary connections are made with strong copper wires (see Poggendorff's *Annalen* 1839, Vol. 48, p. 300; 1840, Vol. 49, p. 511).[306] The cost of the platinum plates has hitherto limited the use of these otherwise powerful and comfortable piles; therefore it will be pleasant for those who, for this reason, could not obtain these plates, that Professor Poggendorff has used iron plates instead of platinum plates with almost the same success.

---

[298] [WW41c] with English translation in [WW20]. See also [WW41a] and [WW41b].

[299] Translated by H. Härtel, haertel@astrophysik.uni-kiel.de and `http://www.astrophysik.uni-kiel.de/~hhaertel/index_e.htm`. Edited by A. K. T. Assis. We thank Robert W. Gray and Steffen Kühn for relevant suggestions.

[300] The Notes by A. K. T. Assis are represented by [Note by AKTA:].

[301] [Note by AKTA:] The names of the authors appear only in the first paragraph of this paper.

[302] [Note by AKTA:] Friedrich Wöhler (1800-1882).

[303] [Note by AKTA:] Johann Christian Poggendorff (1796-1877) edited the *Annalen der Physik und Chemie* from 1824 to 1876, where many of Weber's papers were published. The modern *Annalen der Physik* is the successor to this Journal.

[304] [Note by AKTA:] See also [Pog40].

[305] [Note by AKTA:] The Grove voltaic cell, element, battery or pile was named after its inventor, William Robert Grove (1811-1896), [Gro39].

[306] [Note by AKTA:] [Gro39] and [Sch40].

"Now I am concerned," writes Professor Poggendorff on May 1 of this year, "the cells with two liquids that seem to deserve the most attention and are still not much examined. I have made about 50 such cells and found that almost all of them have the invaluable advantage of giving a constant current, so you can make accurate measurements... I want to tell you only one thing of practical use, namely that in the Grove-cell you can replace the expensive platinum with iron, steel or cast iron as long as you take *concentrated fuming acid* (*acidum nitricum fumans*) instead of the usual nitric acid. You can even dilute this fuming acid with $1\frac{1}{4}$ part of ordinary nitric acid, or to the extent that the iron is not yet attacked. The latter is necessary; if you take the acid too weak, the iron will be violently attacked. In the acid of the specified concentration, the iron remains as bright as the platinum. Here are the elements of the aforementioned cells for smoking concentrated nitric acid and sulphuric acid with 4 parts of water. The zinc was amalgamated."

	Electromotive force	Resistance
zinc and platinum	100.00	13.120
zinc and iron	78.62	11.275
zinc and steel	86.99	12.927
zinc and cast iron	89.63	12.913

"From the resistance come here 4.36 (inch nickel silver wire with 1/6 line diameter) on the closing wire."[307]

"So you see, with the *same* plate size you can get 9/10 of the effect of the Grove-cell with iron. The missing tenth can easily be replaced by enlarging the plates. Incidentally, the *current* is just as *constant* as with the Grove-cell."

In response to the above notification, Messrs. Wöhler and Weber immediately made several attempts to confirm the given information, which at the same time yielded the curious result that a very strong current is produced if you immerse only iron in *both* liquids and also interchange the amalgamated zinc plate dipped in dilute sulphuric acid with an iron plate. This latter plate, because it cannot be amalgamated, was attacked by the sulphuric acid under a weak development of hydrogen gas. This, however, did not impair the effect; rather, it turned out that the effect of this cell was just as constant as Grove's cell. This cell, composed *just of iron* and two liquids, that gives such powerful effects, is of interest for the theory of the pile in general and for the study of the galvanic properties of iron in particular. Cells have already many times been put together in which two identical metals are combined with two different liquids, for instance by Becquerel and De la Rive, of which Fechner gives a list in the *Repertorium der Experimentalphysik* (*Repertory of Experimental Physics*), p. 454 and following;[308] but it seems that only the existence and direction of the current attracted interest, but the further use and investigation were prevented by the weakness and inconsistency of the effect. Such a strong and constant effect as the one described [here], is new. It makes these types of cells really useful, capable of close examination, and deserve special attention. Two pairs, where each plate was only about 3 square inches in surface area, caused thin platinum wires to glow and were sufficient to vigorously decompose

[307][Note by AKTA:] In German: Vom Widerstand kommen hier 4,36 (Zoll Neusilberdraht von 1/6 Linie Durchmesser) auf den Schliessungsdraht.

[308][Note by AKTA:] [Bec29, pp. 14-18], [dlR29, p. 102] and [Fec32, pp. 454-455].

water. Certainly, this subject deserves to be pursued further, unless Professor Poggendorff has perhaps already extended his much more extensive investigation to this topic.

The weak development of hydrogen gas at the iron plates immersed in the dilute sulphuric acid can be easily avoided by using a *tinned iron sheet*, which in this regard does the same service as amalgamated zinc; it even seems to be preferable to the latter because it is thin and durable, while the zinc becomes brittle due to mercury and easily loses part of its amalgam, which, as a gray powder, covers the surface of the plate or settles in the acid, thus weakening the effect of the pile.

Chapter 15

[Weber, 1841a] Measurement of Strong Galvanic Currents with Low Resistance According to Absolute Measure

Wilhelm Weber[309,310,311,312]

It has happened several times in the article on magnetic friction[313,314] that it was important to determine the intensity of a galvanic current based on an absolute measure in order to be able to compare it with the intensity of other currents under any conditions. For instance, when an iron wheel was magnetized by a galvanic current to measure its magnetic friction, it was of interest to measure more closely the current that had produced this effect. For this purpose, the procedure that Faraday has indicated in the seventh series of his experimental investigations on electricity (*Philosophical Transactions* for 1834 and Poggendorff's *Annalen* 1834, Vol. 33, pp. 316 ff.),[315] could have easily been used. In this procedure the strength of the current is measured by the amount of water it decomposes in a given time. However, if the current had been passed through a water decomposition apparatus, it would have been significantly weakened, which should not be the case in those attempts which required an undiminished current.

It happens quite often that the measurement of the absolute current intensity by the amount of decomposed water is not permissible because of the necessary conduction of the current through a water decomposition apparatus. This is especially the case when simple

309[Web41c] with English translation in [Web20a].

310Translated by H. Härtel, haertel@astrophysik.uni-kiel.de and `http://www.astrophysik.uni-kiel.de/~hhaertel/index_e.htm`. Edited by A. K. T. Assis. We thank Robert W. Gray for relevant suggestions.

311The Notes by H. Weber, the editor of the third volume of Weber's *Werke*, are represented by [Note by HW:]; the Notes by Wilhelm Weber are represented by [Note by WW:]; while the Notes by A. K. T. Assis are represented by [Note by AKTA:].

312[Note by AKTA:] The importance of this paper has been discussed by F. Kohlrausch and K. H. Wiederkehr, [WK68].

313[Note by HW:] Wilhelm Weber's *Werke*, Vol. II, p. 200.

314[Note by AKTA:] [Web41g].

315[Note by AKTA:] [Far34b] with German translation in [Far34a].

circuits are used, where a very strong current through that apparatus is so weakened that there is no water decomposition and therefore there can be no question of measuring the decomposed water. In such cases a different method must be used, where the current is only passed through thick[316] and short copper wires, which do not increase the resistance noticeably.

Instead of the method specified by Faraday, the following very simple method was used for the above experiments where a certain piece of the thick current carrying wire was passed in a straight line at some distance from a magnetic needle, so that the latter deviated considerably from the magnetic meridian, while the rest of the circuit was positioned in such a large distance to the needle that its effects did not need to be considered. From the measured deflection of the needle and taking into account the length and position of the active conducting wire as well as the absolute intensity of the Earth's magnetism at the point of observation, an absolute determination of the intensity of the galvanic current is possible — as has been given on p. 49 of the *Resultaten des magnetischen Vereins im Jahre 1840*.[317,318] Incidentally, this method has the advantage that it allows for the determination of the absolute current intensity for every moment, while when using Faraday's method only mean results are obtained for longer periods of time. Experiments can also be carried out where the intensity of one and the same current is measured simultaneously according to this and Faraday's method, and thereby a comparison of the units on which both methods are based would be possible; this comparison, however, is not necessary for the absolute determination of the current intensity. Such a comparison is only necessary when using an ordinary galvanometer, which consists of a magnetic needle provided with a multiplier and which is not directly suitable to obtain absolute determinations, as Jacobi did in Poggendorf's *Annalen*, Vol. 48.[319]

Given the frequently occurring need to determine the absolute intensity of galvanic currents in simple circuits, and with Faraday's method failing, an instrument which is constructed according to the principles mentioned above and which leads directly to the goal can be of great use, which is why some explanations are given here about its most advantageous facility and about some measurements made with it.

The instrument is constructed more appropriately, the greater the distance of the current carrying wire compared to the needle length, because then the distribution of the magnetism in the needle does not need much consideration; all this under the assumption that this greater distance leads to a deflection of the needle which can be observed with sufficient accuracy. From this the advantage is self-evident if the current carrying wire, instead of being guided in a straight line past the needle (which was just done in the above-mentioned attempts in the absence of a proper instrument and merely for the sake of easier execution), it will be guided in a wide vertically directed circle all around the needle. With the same deflection, the distance of all parts of the current carrying wire can then be much larger.[320] In addition, if the conductor forms exactly a vertical circle around the center of the needle, it becomes very simple and easy to calculate the absolute intensity of the galvanic current from the observed deflection of the needle. This circular shape of the conductor has finally the

[316][Note by AKTA:] In German: *Starke*.

[317][Note by HW:] Wilhelm Weber's *Werke*, Vol. II, pp. 202 and 203.

[318][Note by AKTA:] [Web41g, pp. 202-203 of Weber's *Werke*].

[319][Note by AKTA:] [Jac39].

[320][Note by AKTA:] That is, the distance of any portion of the circular wire to the needle can be much larger than the distance of the straight wire to the needle in the previous case.

special advantage that the rest of the conductor can be positioned very easily so that it has no noticeable influence on the needle. For this [purpose,] it is only necessary to put the two conductors, which feed the current in and out, very close to each other, where their effects on the needle mutually cancel. The first pieces of the conducting wires, starting at the ring, should best be conducted through two copper tubes, one enclosing the other but isolated from each other, as shown in Figures 1 to 4. The cross section of the circular conductor must be so large that its resistance is imperceptible.

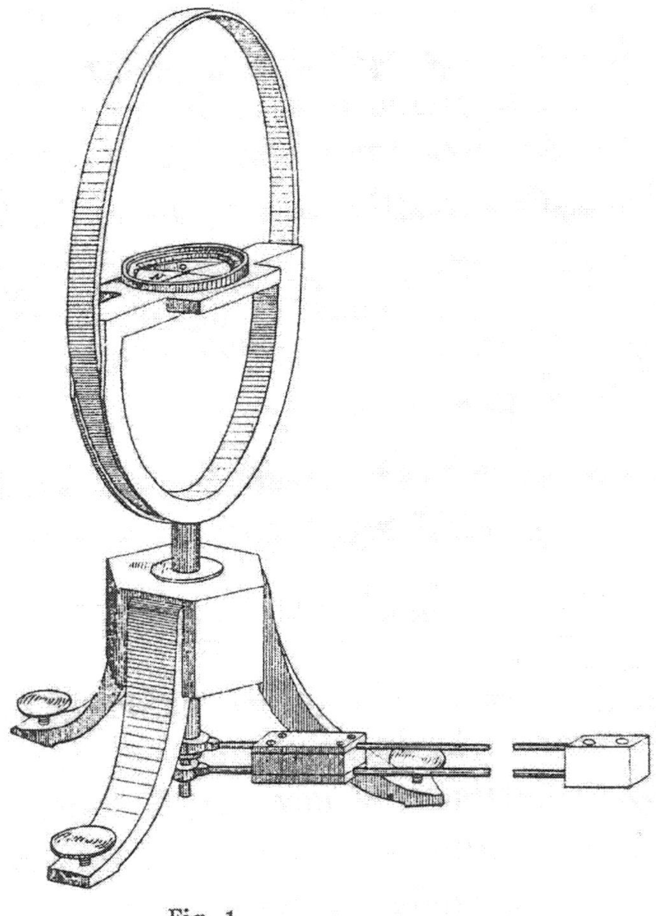

Fig. 1.

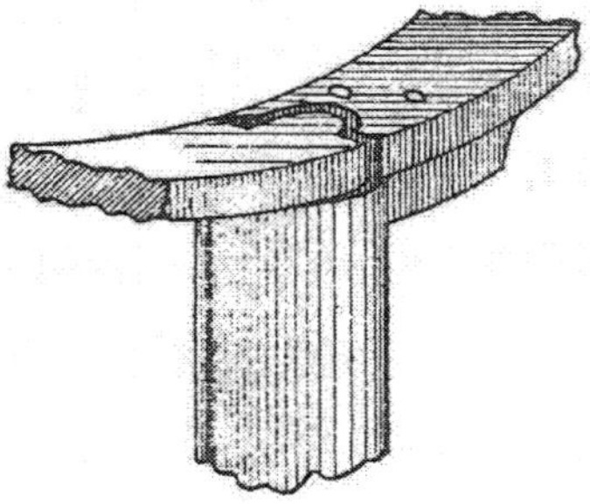

Fig. 2.

Fig. 3.

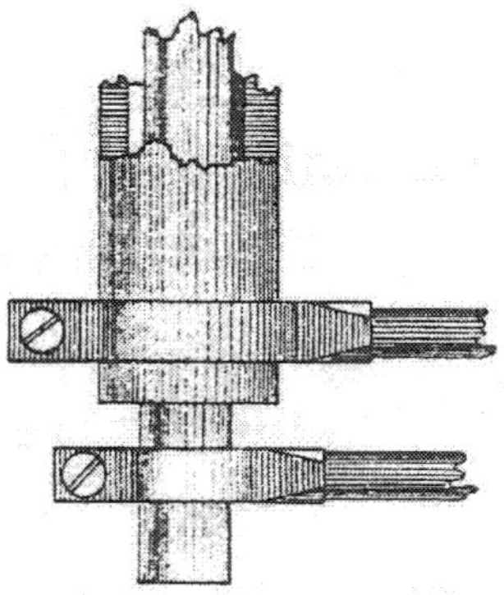

Fig. 4.

I had an instrument set up following this description, where the copper ring was $198\frac{1}{2}$ millimeter in diameter and the cross-section of which was 30 square millimeters. This ring was cut open at the bottom and one end was soldered to the feeding-in and the other end to the feeding-out conductor. These conductors, plugged into tubes, one plucked into another but insulated, led the current 100 millimeter down to two 4 millimeter thick, 1 meter long conducting wires, which run close to each other to two mercury cups, which could be connected to the two plates of the galvanic circuit. The magnetic needle stood in the middle of the circle on a wooden plate attached to the circle. The circle itself stood on a wooden tripod with set screws. The length of the needle was 50 millimeter and it moved on a circular arc divided into degrees. The use of the instrument requires no explanation. The calculation of the absolute intensity from the observed deflection of the needle consists in multiplying the tangent of the deflection angle by a constant number which is derived from the size of the copper ring and from the absolute horizontal intensity of the Earth's magnetism at the observation site. If R $(= 99.125\ mm)$ denotes the radius of the ring, T $(= 1.7833)$ the horizontal intensity of the Earth's magnetism (in Göttingen),[321] then that constant factor is:

$$\frac{1}{\pi} \cdot RT = 56.2675 \ .$$

If φ denotes the observed deflection, then the desired absolute intensity of the measured current is:

$$\frac{1}{\pi} \cdot RT \cdot \tan\varphi = 56.2675 \cdot \tan\varphi \ .$$

For more convenient use, a Table can easily be set up, which shows directly the desired value of the absolute current intensity for each observed value of φ. One will not be able to carry out such absolute measurements as easily and quickly with any other instrument as with this one.

One word remains to be said about the unit of measure of the current intensity on which this calculation is based. That current is assumed to be the unit of measure if, by flowing around the unit area, acts in the same way as the unit of free magnetism which is defined in the *Intensitatis vis magneticae*.[322,323,324,325]

It should also be noted that the observations are greatly facilitated if the compass is provided with a damper, which causes it to come to rest quickly. For more precise measurements it would be necessary to exchange the compass with a small magnetometer, but a much larger copper circle would have to be used, even if the needle was very short, and would only be for instance 60 to 80 millimeter long. The deflection of the needle when measuring strong currents would then still be measurable if the copper ring would also be 600 millimeter in diameter.

Some measurements made with this instrument may now be cited. To assess the greatest effects that can be produced with galvanic currents, it is important to measure the current intensities of the simple circuits without noticeably increasing the resistance they have due to the conductive wires. This measurement then gives directly the maximum of the current intensity, which can be approached by increasing the number of plate pairs in case the

321[Note by AKTA:] As first measured by Gauss, with Weber's help, in his work on the intensity of the Earth's magnetic force reduced to absolute measure, see Chapters 5, 6 and 7.

322[Note by AKTA:] See Chapters 5, 6 and 7.

323[Note by WW:] Note that this current is half as strong compared to a current whose measurement (see page 49 above) was based on a current exerting the unit of torque [or moment of force] on the magnetic needle, given the unit of length of the conductor and [the unit] of distance from the magnetic needle, and [also] given the unit of free magnetism in the needle. This arises easily from the basic law of galvanism, as stated in Section 1 of the *Allgemeine Lehrsätze* in the previous volume of the *Resultate* and has already been mentioned here on page 48.

324[Note by HW:] See Wilhelm Weber's *Werke*, Vol. II, pp. 202 and 203; and Gauss' *Werke*, Vol. V, p. 198.

325[Note by AKTA:] [Web41g, pp. 48-49 of the *Resultate aus den Beobachtungen des magnetischen Vereins im Jahre 1840* and pp. 202-203 of Weber's *Werke*]. The *Allgemeine Lehrsätze* mentioned by Weber refers to Gauss' work "Allgemeine Lehrsätze in Beziehung auf die im Verkehrten Verhältnisse des Quadrats der Entfernung wirkenden Anziehungs- und Abstossungs-kräfte", [Gau40, p. 198 of Gauss' *Werke*], with English translation in [Gau43]: "General propositions relating to attractive and repulsive forces acting in the inverse ratio of the square of the distance".

current has to overcome a larger resistance. The following Table shows the results of these measurements for 5 simple circuits of various sizes and compositions:

Indication of the circuit	Observed deflection	Calculated absolute intensity
A.	72°2′	173.52
B.	78°15′	270.52
C.	66°40′	130.44
D.	54°2′	77.54
E.	73°2.5′	184.52

The following should be noted about the size and composition of these circuits:

A was a Daniell's cell,[326] where the copper area touched by the copper vitriol solution was 9 square decimeter. The copper vitriol solution, as well as the water surrounding the amalgamated zinc rod, was mixed with 10 percent sulphuric acid.

B was a Grove's cell.[327] A platinum beaker with a surface area of 1.9 square centimeter was filled with ordinary nitric acid, while a small porous clay pot filled with dilute sulphuric acid stood in the middle and an amalgamated zinc rod was immersed in the latter. The sulphuric acid was mixed with 80 percent water.

C was a cell according to Professor Poggendorf[328] with an iron plate in smoking nitric acid, instead of the platinum plate in ordinary nitric acid from Grove's column. Nitric acid touched the iron plate from both sides, but the total contact area was only 3/4 decimeter. The sulphuric acid surrounding the clay pot and in which an amalgamated zinc cylinder was immersed was diluted with 90 percent water.

D was a cell of the same size and composition as the previous one, with the only difference that the zinc plate of the previous cell immersed in dilute sulphuric acid was also replaced by an iron plate. Attention has already been drawn to the fact that only one metal is needed for the strong currents that arise here (*Göttinger gel. Anz.* 1841, 81. Stück).[329,330]

E finally was a cell according to Professor Bunsen in Marburg.[331] A coal cylinder made of hard coal and cokes, which was permeated with nitric acid, was immersed in dilute sulphuric acid with a surface area of $1\frac{7}{10}$ square decimeter and surrounded by a zinc cylinder at a short distance. The sulphuric acid was diluted with 90 percent water.

The above results are the largest which were obtained when testing several cells of the same construction. Four examples were tested from the first, fourth and fifth type, two examples from the third type and only one example from type two. The greatest difference in these repetitions occurred with the fifth type and was probably due to the often imperfect conduction of the electricity from the coal into the copper wire. The other 3 cells had given about half the current as the one above.

The strongest current among those measured here was obtained in the above experiments with the Grove circuit, whose intensity was found = 270.52. Such a current, if it passed

[326] [Note by AKTA:] The Daniell voltaic cell or element was named after its inventor, John Frederic Daniell (1790-1845).

[327] [Note by AKTA:] See footnote 305 on page 183.

[328] [Note by AKTA:] See footnote 303 on page 183.

[329] [Note by HW:] Wilhelm Weber's *Werke*, Vol. III, p. 4.

[330] [Note by AKTA:] [WW41c, p. 4 of Weber's *Werke*]. See also [WW41a] and [WW41b].

[331] [Note by AKTA:] The Bunsen voltaic cell or element was named after its inventor, Robert Wilhelm Eberhard Bunsen (1811-1899).

unimpeded through water, would decompose 2.536 milligram water every second, or develop about $4\frac{3}{4}$ cubic centimeter of detonating gas,[332] as will be shown in the following article.[333] If such a current encompasses a square meter of area, it exerts just as great magnetic forces at a distance as a very strong steel magnet weighing 676.3 gram (where you can count 400 [absolute] units of measure of magnetism on 1 milligram of steel).

Thin platinum wires are often used to estimate by their glow the intensity of the current. A measurement showed that a clear, daytime glow of a 2/15 millimeter thick platinum wire was produced by a current whose absolute intensity was = 20. In order to get hold of the amount of heat released in such a wire, a $28\frac{1}{2}$ millimeter long piece of that 2/15 millimeter thick platinum wire was passed through 114 grams of distilled water. The heat released in the water by a galvanic current which was conducted through this wire was shared with the surrounding water and could be measured by the temperature increase of the water in which a thermometer was immersed. The same current, which caused the heating of the wire and of the water, was conducted through the copper circle of the galvanometer and deflected the magnetic needle, placed at the center, from the magnetic meridian. The following Table shows the results of such a series of measurements, where the initial temperature of the water was 15° centigrade.

Time	Deflection	Water-Temperature
11′0″	52°30′	21.5
11′30″	52°30′	22.0
13′30″	51°30′	23.0
15′0″	51°30′	24.0
17′0″	52°0′	25.0
19′50″	51°50′	26.0
20′30″	51°20′	27.0
22′30″	51°0′	28.0
24′30″	50°30′	28.5
26′0″	50°10′	29.0
29′0″	49°20′	30.0

The difference x between the initial temperature of the water and the temperature after t minutes can then be determined as:

$$x = 0.95 \cdot t - 0.015 \cdot t^2 ,$$

from which it follows that if the heat development in the wire is proportional to the current intensity,[334] a current whose intensity is = 1 would heat the described platinum wire during 1 minute so that the temperature of 1 gram of water would rise by 1.4° centigrade. If the wire was cut inside the water, the deflection of the needle was zero, to prove that no measurable part of the current was flowing through the water.

It is to be hoped that in experiments with strong galvanic currents, their absolute intensity is always measured and stated in a similar manner as described here, in order to make the

[332][Note by AKTA:] In German: *Knallluft*. See [Pat17, p. 142].

[333][Note by AKTA:] [Web41h] with English translation in [Web20b], see also [Web42d].

[334][Note by AKTA:] As a matter of fact, James Prescott Joule (1818-1889) discovered also in 1841 that the heat which is evolved is proportional to the square of the intensity of the current, multiplied by the resistance of the wire, [Jou41]. A detailed analysis of Joule's paper can be found in [MS20].

results obtained under different conditions comparable by different observers and to be able to check their agreement.

Chapter 16

[Weber, 1841b] On the Electrochemical Equivalent of Water

Wilhelm Weber[335,336,337,338,339]

After Faraday's numerous experiments,[340] there seems to be no doubt that in the event of chemical decomposition by the galvanic current, for each body the decomposed mass of the same to the related quantity of current, i.e. to the amount of electricity passed through the cross-section of the circuit during the decomposition, is in constant proportion no matter how the galvanic current is produced and in what state the decomposed body may be. The other equally important result found by Faraday must be added to this important law saying that chemically equivalent masses of different bodies need the same amount of electricity (equal quantities of current) to decompose them. For example, 9 grams of water and 36.5 grams of hydrochloric acid are chemically equivalent masses and, according to Faraday, need equal amounts of electricity to be decomposed to oxygen and hydrogen gas, and chlorine and hydrogen gas, respectively. If one then speaks of electricity as of a body that combines with other bodies (with the constituents of the decomposed body) according to their chemically determined equivalent relationships, and if one assumes a certain quantity (positive or negative) of electricity as a measure, and then determines the masses of other bodies which combine with it, then Faraday calls the latter electrochemical equivalents to distinguish them from the chemical equivalents to which they are proportional. The chemical and electrochemical equivalents differ only in the various measures on which they are based, namely on the unit of mass[341] of oxygen (or hydrogen) for those, and on the unit of mass

[335][Web41h] with English translation in [Web20b]. See also [Web42d].

[336]Translated by H. Härtel, haertel@astrophysik.uni-kiel.de and `http://www.astrophysik.uni-kiel.de/~hhaertel/index_e.htm`. Edited by A. K. T. Assis. We thank Robert W. Gray for relevant suggestions.

[337]The Notes by H. Weber, the editor of the third volume of Weber's *Werke*, are represented by [Note by HW:], while the Notes by A. K. T. Assis are represented by [Note by AKTA:].

[338][Note by HW:] *Results from the Observations of the Magnetic Association in 1840*, edited by Karl Friedrich Gauss and Wilhelm Weber, Leipzig, 1841, p. 91-98 and *Annalen der Physik und Chemie*, edited by J. C. Poggendorff, Volume 55, Leipzig, 1842, pp. 181-189.

[339][Note by AKTA:] The importance of this paper has been discussed by F. Kohlrausch and K. H. Wiederkehr, [WK68].

[340][Note by AKTA:] Michael Faraday (1791-1867). See [Far33] and [Far34a], with German translations in [Far34c] and [Far34b], respectively.

[341][Note by AKTA:] In German: *Masseneinheit*.

of electricity for these. Faraday himself has not specified the mass of electricity, which he accepts as a unit. However, if one wanted to take the mass which combines with the mass unit oxygen (or hydrogen) in water to form oxygen gas (or hydrogen gas), the two types of equivalent masses would be completely identical. Therefore, if electrochemically equivalent masses are to mean something different than chemically equivalent masses, they must be measured according to another fundamental unit[342] of electricity, which is derived from another class of electrical effects. The class of the magnetic effects of electricity in galvanic currents is the most suitable for this, since these effects have been reduced to absolute measures in the teaching of magnetism and precise measurement methods have been developed for them.

As absolute measure of electricity (positive or negative or both together) is taken hereafter to be the amount of electricity that has to go through the cross-section of a conductor in the time unit (second), which limits the unit area in one plane, to produce in the distance identical effects as the absolute basic unit of free magnetism.[343]

It will now be of particular interest, based on this absolute measure of electricity, to determine the electrochemical equivalent of any body, e.g. that of water, from which it is then easy to derive the electrochemical equivalents of other bodies according to the laws discovered by Faraday with the help of their chemically determined equivalents to which they are proportional. The determination of the electrochemical equivalent of water on the basis of the measure of electricity specified above is now to be the subject of this article.

For this purpose, it is necessary to observe any measurable magnetic effect of the galvanic current while a certain quantity of water is being decomposed. To do so, neither the effect of the current on the Pouillet's sine-galvanometer, nor on the tangent-galvanometer from Nervander can be used, because these instruments though they can give correct comparisons of the current intensities, they cannot give absolute determinations.[344] The instrument described in the previous article[345,346] therefore seems to be the only instrument suitable for this task. Indeed, this is the easiest and most convenient method if it is not about finer measurements, and even these could be carried out with this instrument if it would be used in the finer manner specified above,[347,348] namely with a very large copper circle and a very small needle, hung on a thread, like in a magnetometer, and provided with a mirror so that it can be observed with a telescope and a scale.

In the absence of the finer version of such an instrument, I have used an instrument based on other principles and intended for other purposes, of which I should briefly mention what is necessary for the present purpose. When applying this instrument, no magnetic needle is used, but only the conductor of the galvanic current itself.

A copper wire, isolated with silk and of known length is carefully wound on a cylindrical

---

[342][note by AKTA:] In German: *Grundmaasse.* This expression can also be translated as "basic unit" or "basic dimension".

[343][Note by AKTA:] That is, as a magnetic dipole with a magnetic moment = 1. The axis of this dipole must be considered orthogonal to the current loop.

[344][Note by AKTA:] In German: *Sinus-Boussole* and *Tangenten-Boussole.* The tangent galvanometer was invented by Johan Jakob Nervander (1805-1848) and the sine galvanometer by Claude Servais Mathias Pouillet (1790-1868), [Ner33], [Pou37] and [Sih21]. Friedrich Kohlrausch discussed measurement of currents with the tangent and sine galvanometers, [Koh83, Chapters 64 and 65, pp. 188-192].

[345][Note by HW:] Wilhelm Weber's *Werke*, Vol. III, p. 8.

[346][Note by AKTA:] [Web41c, p. 8 of Weber's *Werke*].

[347][Note by HW:] In the same place, p. 10.

[348][Note by AKTA:] [Web41c, p. 10 of Weber's *Werke*].

coil[349] of a certain diameter, so that all turns come very close to a system of concentric circles. The area of these circles can be equated with the area covered by the wire, which is based on the length of the wire, the diameter of the coil and the number of turns. This can be easily calculated, and will be denoted by S.

The two ends of the wire lead to two small mutually insulated metallic hooks on the coil, to which two other not overspun fine wires are attached, on which the entire coil of wire is suspended in a *bifilar* manner.

The *bifilar* suspension of the coil on the latter two wires has a dual purpose: *firstly*, the same as with the bifilar-magnetometer,[350] in order to obtain a certain directive force[351] D and then to determine all the forces acting on the coil and trying to turn it. Although this directive force could be calculated from the length of the suspension wires, their spacing and the weights they carry (insofar as their own elasticity does not have to be taken into account), the same can be found more precisely through the experiments specified in the *Intensitas*[352] for determining the moment of inertia, which can be referred to here.

Secondly, these two suspension wires have the special purpose that they form the bridge through which the current is fed both from the outside to the wire and back again, without the slightest impairment of the mobility of the coil, as would be the case if you need metal tips attached to the coil and dipped into mercury wells where the inevitable friction does not allow measurements.

The bifilar suspension ensures that even when the current passes through the coil, its position and the vibrations can be observed with the same freedom as the position and the vibrations of the bifilar magnetometer. It is therefore permitted to use the same fine tools for their observation, namely to attach a mirror to the coil and to observe the image of a distant scale with a telescope. In this way, the path to the finest galvanic measurements is paved without using magnetic needles.

It is easy to first set up the tripod on which the coil is suspended so that the coil maintains the same position when a current of any strength is passed through the coil, sometimes forward and sometimes backward, and then to rotate the whole system around a vertical axis by 90°. Then the instrument for the execution of our measurement is prepared.

The measurement then consists in the fact that the same current, which decomposes the water in the water decomposition apparatus, is passed through our instrument, where the force of the horizontal part of the earth's magnetism causes a deflection. This deflection must be observed closely in short intervals during the duration of the water decomposition. It is then easily understood that the absolute intensity G of the galvanic current for any moment in time while the deflection φ is observed is given by the following equation:

$$STG = D \tan \varphi \ ,$$

where T denotes the absolute horizontal intensity of the earth's magnetism at the observation site. If T is known and S and D are exactly determined, as stated above, the intensity G can be calculated from the observed deflection φ. From all their values for the time interval t where the water decomposition occurred, the amount of electricity E that passed through

[349][Note by AKTA:] In German: *Einer cylindrischen Rolle*. The word "Rolle" can also be translated as "roll", "bobbin" or "pulley".

[350][Note by AKTA:] See [Web38a] with English translation in [Web41d] and [Web66b]; see also [Web94c].

[351][Note by AKTA:] See footnotes 246 and 263 on pages 125 and 137.

[352][Note by AKTA:] See Chapters 5, 6 and 7.

the coil and was used for the decomposition of the water can be determined with great accuracy through

$$E = \int G dt \ ,$$

according to the absolute units specified above. If we hereby divide the amount of the decomposed water W expressed in milligrams, the quotient W/E gives the amount of water which is decomposed by the absolute amount of electricity specified, i.e. the desired electrochemical equivalent of water.

After this description of the measurement method used, the results of the measurements can be briefly summarized.

The wire wound on the coil formed 1130 turns; the periphery of the coil was 164 millimeters; the length of the wire 253 600 millimeters. This results in S:

$$S = 4\,638\,330 \text{ square millimeters.}$$

The moment of inertia K of the coil was found according to known regulations:

$$K = 779\,400\,000 \ .$$

The period of oscillation t,[353] which changed somewhat with temperature, was

for the first and second measurement $t = 8.0702''$	118 111 000 ,
for the third and second measurement $t = 8.0803''$	117 817 000 ,
for the forth and fifth measurement $t = 8.0904''$	117 523 000 ,

from which the values of the directive force $\pi^2 K/t^2$ follows,[354] given in the last column.

At the time of these experiments in Göttingen, the absolute horizontal intensity T of the earth's magnetism could be measured after an almost simultaneous measurement in the magnetic observatory with the result[355]

$$T = 1.7833 \ ;$$

however, these observations were not made in an iron-free establishment, but in a room of the astronomical observatory[356] where there was a great deal of iron at moderate distances. The horizontal intensity at this observation point was therefore compared with that in the magnetic observatory by means of comparative measurements, and the absolute intensity of the earth's magnetism for the place where the experiments were carried out resulted in:

$$T = 1.7026 \ .$$

Finally, the simultaneous observation of the water decomposition apparatus and the galvanometer in the five measurements gave the following results:

[353][Note by AKTA:] In German: *Die Schwingusdauer t*. See also footnote 176 on page 58.

[354][Note by AKTA:] Weber is here utilizing the equation of motion of a rigid body as $\tau = -D\varphi = K\ddot{\varphi}$, where K is the moment of inertia of the body, $\tau = -D\varphi$ is the torque or rotational moment acting on it when it suffers a deflection φ, while D is the directive force acting on the body.

[355][Note by AKTA:] As first measured by Gauss with Weber's help in 1832, see Chapters 5, 6 and 7.

[356][Note by AKTA:] In German: *Sternwarte*.

	Decomposed Water in milligrams	Time interval for the decomposition	Amount of electricity based on absolute measures
1.	14.2346	1168"	1522.44
2.	14.2026	1280"	1504.92
3.	14.0872	1137.5"	1506.46
4.	14.0812	1154"	1501.43
5.	13.9625	1263"	1484.90

From this it follows for the electrochemical equivalent of water the following five results:

0.009 350	−0, 000 026
0.009 437	+0,000 061
0.009 351	−0, 000 025
0.009 337	−0, 000 039
0.009 403	+0,000 027

therefore as a mean value 0.009 376.[357]

The differences between the individual measurements from this mean are noted in the last column.

It should be added that the amount of the decomposed water was determined as usual, from the volume of the evolved gases, and both gases were collected and measured. In order to avoid the absorption of the gases by the water, the former was collected over a mercury trough, which Professor Wöhler[358] was kind enough to lend. The water to be decomposed consisted of a few drops, which, mixed with sulfuric acid, occupied the sealed end of an S-shaped tube (and represented the function of a retort). The atmospheric air was completely excluded. Two platinum wires, which were melted into the tube and passed through the water without touching one another, were used to conduct the galvanic current through the water. The decomposition of water had started long before the start of the measurement. The gas was measured [while it was] humid. The walls of the tube in which it was collected had been moistened with distilled water before being filled with mercury. The influence of temperature and barometer readings were also taken into account properly. The observations were all carried out jointly by Professor Ulrich[359] and the undersigned.

As for the result itself, the consistency between the five measurements can be seen as a new confirmation of the Faraday theorem that the same amount of electricity is always needed to decompose the same amount of water. If conditions permit in the future, to make that confirmation even more striking, these measurements will be repeated under even more modified conditions. Similar measurements will also be made for other bodies instead of water, e.g. will be carried out with hydrochloric acid.

If one finally compares the result of these measurements with those of the previous article[360,361] on the maximum of the current intensity of different columns, one obtains, as

[357][Note by AKTA:] From this mean value we obtain $1/0.009376 = 106.655 \approx 106\frac{2}{3}$. In later works Weber will utilize this value $106\frac{2}{3}$ whenever referring to the electrolytic unit of current. See, for instance, [WK56, p. 600 of Weber's *Werke*] with English translation in [WK03, p. 290] and Portuguese translation in [WK08, p. 96]; [KW57, pp. 614, 649 and 650 of Weber's *Werke*] with English translation in [KW21, p. 8]; [Web62, p. 88 of Weber's *Werke*]; and [Web64, p. 165 of Weber's *Werke*] with English translation in [Web21a].

[358][Note by AKTA:] Friedrich Wöhler (1800-1882).

[359][Note by AKTA:] Georg Karl Justus Ulrich (1798-1879).

[360][Note by HW:] Wilhelm Weber's *Werke*, Vol. III, p. 10.

[361][Note by AKTA:] [Web41c, p. 10 of Weber's *Werke*].

already mentioned there, knowledge about the rate of water decomposition which can be achieved with the galvanic current under particularly favourable conditions. From this it can be judged whether the galvanic current is a practical tool for the production of oxygen and hydrogen gas. It requires no further discussion that the collected result are useful in the experiments made with Faraday's Volta-electrometer for determining the absolute amounts of electricity more precisely, and for concluding the magnetic effects that could be produced.

Chapter 17

[Weber, 1842] Measurement of Strong Galvanic Currents According to Absolute Measure

Wilhelm Weber[362,363,364,365]

The instrument[366] which is to be used for this measurement is set up in such a way that the current remains noticeably the same, it may or may not be passed through the instrument for the purpose of the measurement. This important point of measuring the current without decreasing it, when it is used, is achieved by making the resistance of the measuring device negligible compared to the remaining resistance of the circuit.

The instrument therefore consists, as Figure 1 shows, of a single thick copper ring,[367] which is set up in the plane of the magnetic meridian, and in its axis a small magnetic needle is located (the length of which is only about the fourth part of the ring diameter). The supply and removal of the current[368] is set up in such a way that only the current which goes through the ring can act on the needle, as can be easily understood from the illustration in Figures 1, 2, 3 and 4.[369]

---

[362][Web42b] with English translation in [Web20c].

[363]Translated by H. Härtel, haertel@astrophysik.uni-kiel.de and `http://www.astrophysik.uni-kiel.de/~hhaertel/index_e.htm`. Edited by A. K. T. Assis. We thank Robert W. Gray for relevant suggestions.

[364]The Notes by H. Weber, the editor of the third volume of Weber's *Werke*, are represented by [Note by HW:]; the Notes by Wilhelm Weber are represented by [Note by WW:]; while the Notes by A. K. T. Assis are represented by [Note by AKTA:].

[365][Note by AKTA:] The importance of this paper has been discussed by F. Kohlrausch and K. H. Wiederkehr, [WK68].

[366][Note by AKTA:] Prof. Karin Reich called my attention of a paper by Weber dealing with this instrument which is not contained in Weber's *Werke*: [Web42c].

[367][Note by AKTA:] In German: *aus einem einzigen starken Kupferrringe.*

[368][Note by AKTA:] In German: *Die Zuleitung und Ableitung des Stroms.*

[369][Note by AKTA:] Further details of this instrument and its utilization can be found in [Web41c] with English translation in [Web20a].

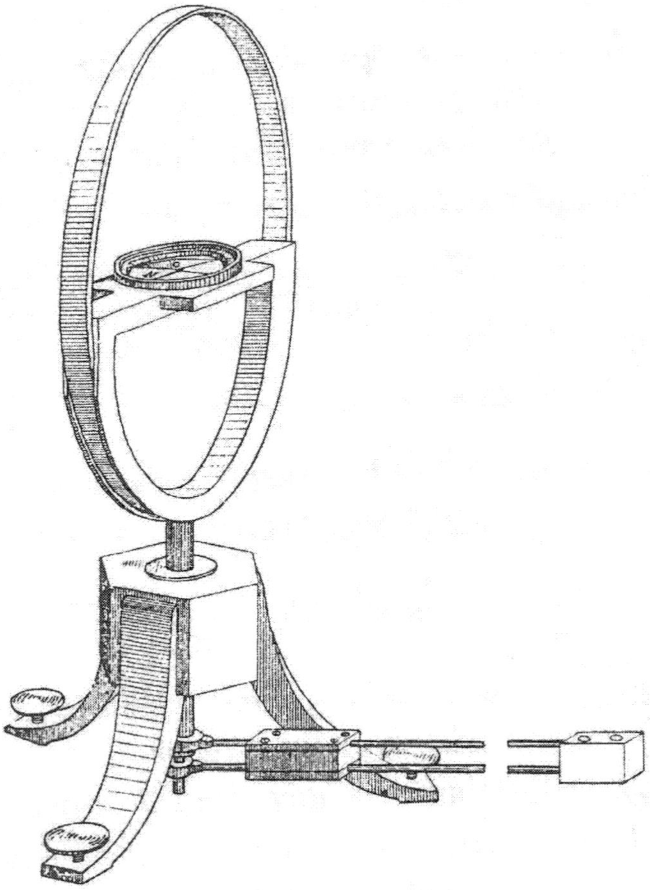

Fig. 1.

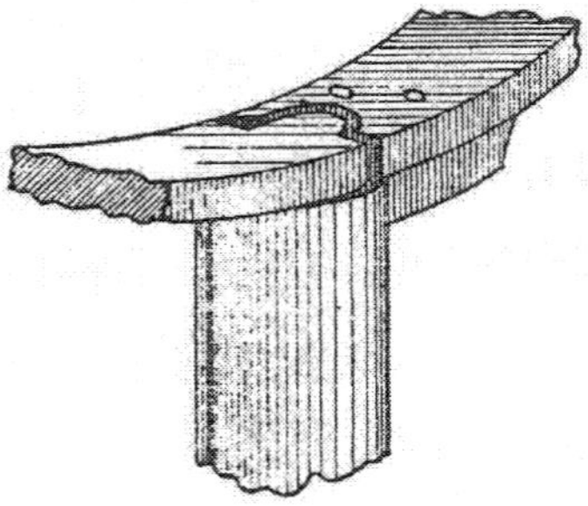

Fig. 2.

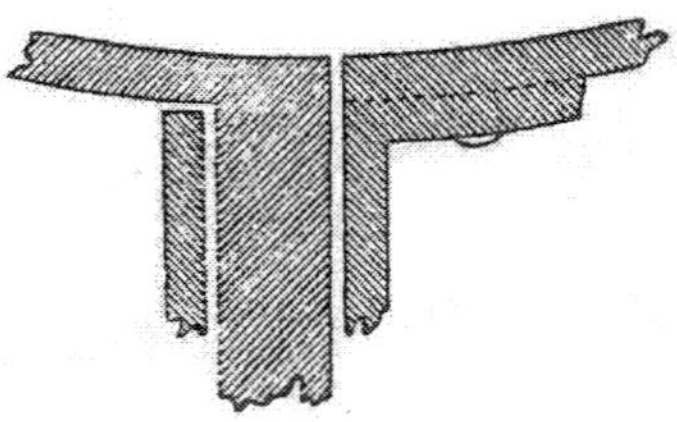

Fig. 3.

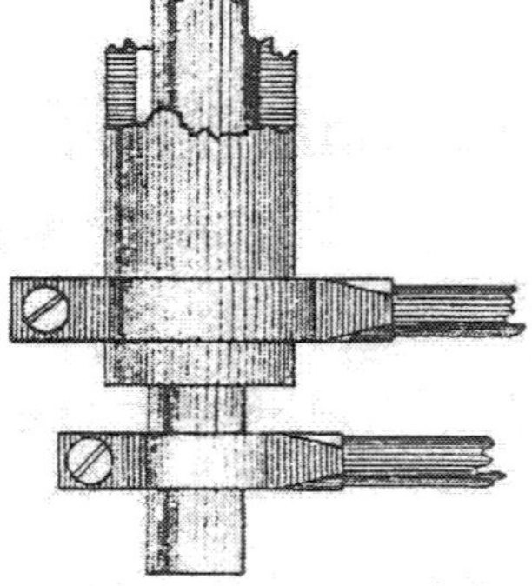

Fig. 4.

This setup of the instrument does not need any further explanation, since it essentially corresponds to the setup of a *tangent galvanometer*,[370] as it has already been used frequently. It should therefore only be shown in more detail how one can use it to determine the strength of a galvanic current according to *absolute measure*, which is easily accomplished if one assumes as known the Gaussian method of measuring magnetism according to absolute measure (see Ann., Vol. 28, pp. 241, 591).[371]

Just like the *moment of a magnet*, the *moment of a closed galvanic circuit* can also be measured in *absolute units* (if the Earth's magnetism is known) from the deflection of a magnetic needle from the magnetic meridian caused by the current. Essentially the same rules must be observed here as there in order to obtain a safe and precise result.

If the *moment of a magnet* is to be measured, the deflection of a needle is observed at two different distances from the magnet. It is assumed that the magnet is always positioned in the horizontal plane of the needle and perpendicular to the magnetic meridian, and that its extended axis meets the centre of the needle. The deflection should be observed 4 times at every distance, where the magnet should be positioned now to the East, now to the West of the needle and its North pole should turn now to the East, now to the West. Let the

[370][Note by AKTA:] See footnote 344 on page 196.

[371][Note by AKTA:] Weber is quoting here Gauss' work on the intensity of the Earth's magnetic force reduced to absolute measure, [Gau33b]. See Chapters 5, 6 and 7.

mean values of the deflection found for the distances R and R' be v and v'. Now put:

$$\tan v = \frac{L}{R^3} + \frac{L'}{R^5} ,$$

$$\tan v' = \frac{L}{R'^3} + \frac{L'}{R'^5} .$$

This can be accepted if R and R' are so large compared to the length of the magnet and the needle, that the terms of the series which contain the 7th or higher order of R and R' can be neglected. By eliminating L' one obtains from this:

$$L = \frac{R^5 \tan v - R'^5 \tan v'}{R^2 - R'^2} ,$$

where v, v', R and R' are known by measurements. The theory has proven that the following relation exists between the calculated value of L and the desired moment M of the magnet:

$$L = \frac{2M}{T} \qquad \text{or} \qquad M = \frac{1}{2}LT ,$$

where T denotes the horizontal intensity of the Earth's magnetism in absolute measure.

This method, presumed to be known, of measuring the *moment of a bar magnet* according to absolute measure, could be applied directly to the measurement of the *moment of a closed galvanic circuit*, if this complete circuit did not occupy a larger space than the magnet, and thereby would cause at the same distance an equally large deflection of the needle. But since these two conditions cannot be met at the same time, the following modification of the method can be accepted when it is applied to galvanic circuits.

The galvanic current is passed through a large and thick copper ring positioned in the plane of the magnetic meridian. The supply of the current to the ring is done through a long and thick copper rod, the discharge of the current is through a copper tube that surrounds the rod without touching it. The magnetic needle is set up in such a way that it is at equal distance from all parts of the ring; the centre of the needle lies in the axis of the ring either at its centre or near to it, so that the current flows almost all the way around the needle.

Let the point A in Figure 5 be the centre of the ring, AB its axis, $AC = y$ its radius; the intensity of the current is called g.

Fig. 5.

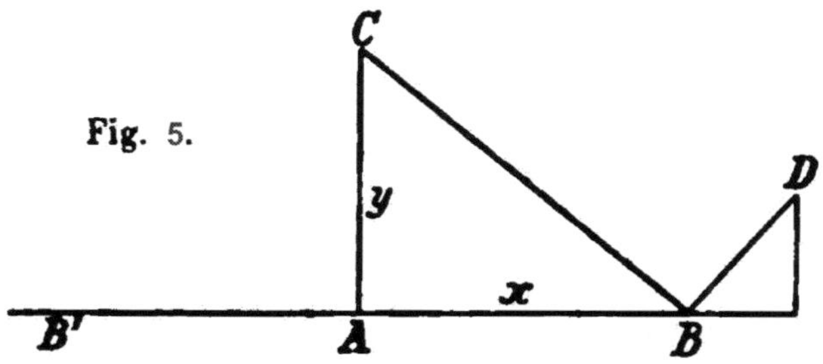

Let a North magnetic element μ be located on the axis at the distance $AB = x$ from the centre. If the current g passes through the ring element $yd\varphi$ at point C (from back to front in the Figure), μ will be moved from B to D and this motion will be perpendicular to that plane through B and through the ring element at C. The magnitude of this moving force is directly proportional to the product $g\mu y d\varphi$ and inversely proportional to the square (x^2+y^2) of the distance CB,[372] or it can be expressed through

$$\frac{fg\mu y d\varphi}{x^2+y^2},$$

where f denotes a constant factor. If one decomposes this force BD along the direction of the ring axis by multiplying that value by $y/\sqrt{x^2+y^2}$, which gives $fg\mu y^2 d\varphi/(x^2+y^2)^{3/2}$, the resultant of the force is obtained with which all elements $yd\varphi$ of the circular current try to move the element μ in the direction of the axis, [namely:]

$$=\frac{2\pi fg\mu y^2}{(x^2+y^2)^{3/2}}.$$

The forces perpendicular to the direction of the axis cancel each other.

If one compares this force with that which an infinitely small *magnet*, whose axis coincides with the direction AB and whose moment is M, at the distance $CB=\sqrt{x^2+y^2}$ from the element μ located in B, would exert, [namely:]

$$=\frac{2M\mu}{(x^2+y^2)^{3/2}},$$

(see Gauss in the *Resultaten des magnetischen Vereins für das Jahr 1840*, p. 26 and the following, and the article: "Bemerkungen über die Wirkungen eines Magnetes in die Ferne" (Remarks on the actions of a magnet at a distance),[373,374] so you can see that both expressions become identical for

$$M=\pi fgy^2 .$$

If, in analogy with the magnetic moment, πfgy^2 is called the *moment of the galvanic circular current*, and denoting it with G, then G can be determined by deflection experiments according to absolute measure, just like M. If one denotes by u the observed deflection[375] of a magnetic needle in A, with u' the observed mean deflection of the same in B and B' (where $B'A=BA$), and sets $y=R$, $\sqrt{x^2+y^2}=R'$, it follows in a similar way:

$$\tan u=\frac{L}{R^3}+\frac{L'}{R^5},$$

$$\tan u'=\frac{L}{R'^3}+\frac{L'}{R'^5}.$$

It follows by elimination of L':

$$L=\frac{R^5\tan u-R'^5\tan u'}{R^2-R'^2}=\frac{2G}{T},$$

or

---

[372][Note by WW:] The sine of the angle that CB makes with the direction of the ring element in C should be added as a factor, which in our case is equal to 1 because that angle is a right one.

[373][Note by HW:] Gauss' *Werke*, Vol. V, p. 427 and Wilhelm Weber's *Werke*, Vol. II, p. 242.

[374][Note by AKTA:] [Gau41d, pp. 26 and the following of the *Resultate* and pp. 427 and the following of Gauss' *Werke*] and [Web42a].

[375][Note by WW:] It is assumed here that each deflection observation is repeated after the current in the ring has been reversed.

$$G = \frac{1}{2} \frac{R^5 \tan u - R'^5 \tan u'}{R^2 - R'^2} \cdot T = \pi f g R^2 \ .$$

From this one finally finds the searched *intensity* g of the current when one determines the unit on which its determination is based. If one takes that current intensity as the unit,[376] *whereby the current, when it circulates in the plane the unit of an area, exerts the same action at a distance as the unit of free magnetism*, then the indeterminate factor f is determined, because then at the same time the intensity $g = 1$, the moment $G = 1$ and the area $\pi R^2 = 1$, from which the value of f results:

$$f = 1 \ ;$$

it follows

$$g = \frac{LT}{2\pi R^2} \ ,$$

where L can be calculated from the measured quantities u, u', R, R'.

This determination of the absolute intensity of the galvanic current becomes even easier if the length of the needle can be neglected compared to the diameter of the circle, because one can then restrict oneself to the first term in the series expansion for $\tan u$:

$$\tan u = \frac{L}{R^3} \qquad \text{or} \qquad L = R^3 \tan u \ .$$

One then only needs to measure the deflection u when the needle is in the center of the circle and then get

$$g = \frac{1}{2\pi} RT \tan u \ .$$

This approximation formula can still be regarded as sufficient even for measurements of higher quality, if the length of the needle does not exceed the fourth or fifth part of the diameter, as one can convince oneself if one carries out the observations completely, as stated above, and then compares the results of the approximation formula with the result of the more precise calculation.

The accuracy of the result finally depends on the accuracy with which the deflection u is measured. If the error du is made in this measurement, an error is caused in the current intensity calculated from it, which is in parts of the total intensity $= 2du/\sin 2u$. This error has a minimum for $u = 45°$. From this follows the rule for the construction of the instrument that the copper ring is of the most advantageous size when the current to be measured produces a deflection of 45°, which is the case only with strong currents.

[376] [Note by AKTA:] That is, $g = 1$.

Chapter 18

Gauss and Weber's Absolute System of Units and Its Difference to the Modern "Gaussian" System of Units

A. K. T. Assis[377]

18.1 The Absolute System of Units of Gauss and Weber

In December 1832 C. F. Gauss announced a paper on the intensity of the Earth's magnetic force reduced to absolute measure to the Royal Society of Sciences in Göttingen. The original paper in Latin was published only in 1841, although a preprint appeared already in 1833 in small edition. Several translations have been published since 1833, see Chapters 5, 6 and 7.

In this work Gauss introduced the so-called absolute system of units. Wilhelm Weber began to work with Gauss in Göttingen in 1831, when he was hired as Professor of physics of the University. Gauss mentioned in his 1833 paper that he was assisted by Weber in many ways in his experiments.

It is difficult to maintain unaltered over long periods of time the properties of any magnet, like its degree of magnetization. The same can be said of the magnetic properties of any standard or even of the Earth itself. In order to circumvent this problem, Gauss proposed to base the definition and measurement of the magnetic properties of magnets and of the Earth on the mechanical standards and units of measurement of length, mass and time. This approach is the essence of the absolute system of units. Let us see how this system works for magnetic poles and electric charges.

To Gauss the basic mechanical units of length, mass and time were the millimeter (mm), the milligram (mg) and the second (s):[378]

> If we take the second, the millimeter and the milligram for the units of time, distance and mass, [...]
>
> [...]

[377]Homepage: `www.ifi.unicamp.br/~assis`
[378][Gau03, Section 11, p. 14 and Section 22, p. 25].

[...] millimeters, milligrams, and the seconds of the mean sun time are taken as units, [...]

In this work Gauss also discussed how to convert these units into other units.

Weber also utilized in his works these three basic mechanical units of millimeter, milligram and second.[379]

As regards the force law, Gauss[380] and Weber[381] utilized Newton's second law of motion[382] in the form

$$f = \frac{d(mv)}{dt} = ma \; . \tag{18.1}$$

Here f is the net force acting on a mass m moving with velocity v relative to an inertial frame of reference, t is time and $a = dv/dt$ is the acceleration of the mass m relative to this inertial frame of reference.

Therefore, one unit of force acting constantly on one unit of mass ($1\ mg$) would produce one unit of acceleration ($1\ mm/s^2$), that is, its velocity would change one unit in one unit of time:

$$f = 1\ mg \times 1\frac{mm}{s^2} = 1\frac{mg \cdot mm}{s^2} \; . \tag{18.2}$$

In the modern International System of Units MKSA (based on the meter (m), kilogram (kg), second (s) and Ampère (A)), this absolute unit of force can be expressed as:

$$f = 1\frac{mg \cdot mm}{s^2} = 1 \times 10^{-9}\ \frac{kg \cdot m}{s^2} = 1 \times 10^{-9}\ N \; . \tag{18.3}$$

That is, one absolute unit of force is equivalent to 10^{-9} newtons.

Isaac Newton in 1687 expressed his gravitational force between two particles with masses m_1 and m_2 separated by a distance r saying in words that it was proportional to $m_1 m_2/r^2$. Coulomb in 1785 expressed his electric force between two particles electrified with charges q_1 and q_2 separated by a distance r saying in words that it was proportional to $q_1 q_2/r^2$. Analogously, he expressed his force between two magnetic centers of force (with pole strengths p_1 and p_2) separated by a distance r saying that it was proportional to $p_1 p_2/r^2$.

Gauss[383] and Weber[384] changed this procedure and utilized equalities instead of proportionalities. The gravitational, electric and magnetic forces f were then written as, respectively:

$$f = \frac{m_1 m_2}{r^2} \; , \tag{18.4}$$

$$f = \frac{q_1 q_2}{r^2} \; , \tag{18.5}$$

$$f = \frac{p_1 p_2}{r^2} \; . \tag{18.6}$$

[379] See, for instance, pages 65, 297, 593, 600, 607, 618, 648, 650, 653, 656 and 662 of Volume 3 of Weber's *Werke*, [Web93b].

[380] [Gau03, Sections 1 and 26].

[381] [Web93b, pp. 297, 600, 607, 618, 648, 650, 653, 656 and 662].

[382] See footnote 28 on page 15.

[383] [Gau03, Sections 1 and 26] and [Gau43, Section 1, p. 153].

[384] [Web93b, pp. 157, 237, 244, 366, 667-668].

Here I will consider specifically the electric charge, although similar reasoning might be applied to the gravitational mass and to the magnetic pole. One unit electric charge is that amount concentrated on a particle which exerts one unit of force ($1\ mg \cdot mm/s^2$) in an equal amount of electric charge concentrated on another particle separated from the first one by one unit of distance ($1\ mm$). Let us express this unit electric charge by $q_1 = q_2 = q$. From Equation (18.5) one obtains:

$$q = r\sqrt{f} = 1\ mm\sqrt{1\frac{mg \cdot mm}{s^2}} = 1\frac{mm^{3/2}mg^{1/2}}{s} . \tag{18.7}$$

Weber and R. Kohlrausch, for instance, expressed equations (18.5) and (18.7) in words as follows:[385]

> The unit of electrical fluid is determined in electrostatics by means of the *force*, with which the free electricities act on each other at a distance. If one imagines two equal amounts of electricity of the same kind concentrated at two points, whose distance is the unit of length, and if the force with which they act on each other repulsively, is equal to the *unit of force*, then the amount of electricity found in each of the two points is the *measure* or the *unit* of free electricity

Coulomb's law in the International System of Units MKSA is written as:

$$f = \frac{q_1 q_2}{4\pi\varepsilon_o}\frac{1}{r^2} . \tag{18.8}$$

Here the force f is expressed in newtons, N, the charges q_1 and q_2 are expressed in coulombs, C, the distance r is expressed in meters, m, while $\varepsilon_o = 8.85 \times 10^{-12}\ C^2N^{-1}m^{-2}$ is the so-called permittivity of free space.

How many coulombs are there in 1 absolute unit of charge? The simplest way to answer this question is to consider two equal charges $q_1 = q_2 = q$ separated by 1 absolute unit of distance, $r = 1\ mm = 10^{-3}\ m$, exerting on one another 1 absolute unit of force, $f = 1\ mg \cdot mm/s^2 = 1 \times 10^{-9}\ N$. Equation (18.8) then yields for this case:

$$q = 3.3 \times 10^{-13}\ C . \tag{18.9}$$

That is, a charge of $3.3 \times 10^{-13}\ C$ is equivalent to 1 absolute unit of charge $q = 1\ mm^{3/2}mg^{1/2}s^{-1}$.

18.2 The CGS System of Units

In 1873 the British Committee for the Selection and Nomenclature of Dynamical and Electrical Units decided to adopt the absolute system of units introduced by Gauss and Weber. This committee was composed by W. Thomson (Lord Kelvin), G. C. Foster, J. C. Maxwell, G. J. Stoney, F. Jenkin, Siemens, F. J. Bramwell and Everett. However, through the suggestion of W. Thomson, it changed the basic units of mass, length and time. Instead of Gauss and Weber's millimeter, milligram and seconds, they adopted the centimeter, gram and second. The initial letters of these units gave rise to the so-called CGS system of units:[386]

[385] [WK56, p. 11] and [WK03, p. 288].
[386] [Tho74].

We accordingly recommend the general adoption of the *Centimetre*, the *Gramme*, and the *Second* as the three fundamental units; and until such time as special names shall be appropriated to the units of electrical and magnetic magnitude hence derived, we recommend that they be distinguished from "absolute" units otherwise derived, by the letters "C.G.S." prefixed, these being the initial letters of the names of the three fundamental units.

As regards the unit of force, they introduced the *dyne*, namely:[387]

As regards the name to be given to the C. G. S. *unit of force*, we recommend that it be a derivative of the Greek $\delta\acute{\upsilon}\nu\alpha\mu\iota\varsigma$. The form *dynamy* appears to be the most satisfactory to etymologists. *Dynam* is equally intelligible, but awkward in sound to English ears. The shorter form, *dyne*, though not fashioned according to strict rules of etymology, will probably be generally preferred in this country. Bearing in mind that it is desirable to construct a system with a view to its becoming international, we think that the termination of the word should for the present be left an open question. But we would earnestly request that, whichever form of the word be employed, its meaning be strictly limited to the unit of force of the C. G. S. system—that is to say, *the force which, acting upon a gramme of matter for a second, generated a velocity of a centimetre per second.*

Therefore, expressing one unit of force of the CGS system in the International System of Units MKSA yields:

$$f = 1g \times 1\frac{cm}{s^2} = 1\ dyne = 1 \times 10^{-5}\frac{kg \cdot m}{s^2} = 1 \times 10^{-5}\ N\ . \tag{18.10}$$

That is, while one unit of force in Gauss and Weber's absolute system is equivalent to $10^{-9}\ N$, one unit of force in the CGS system is equivalent to $10^{-5}\ N$.

In 1881 the First International Electrical Congress held in Paris resolved to endorse these three fundamental units of the CGS system. Among the participants of this Congress were Clausius, Helmholtz, William Thomson (Kelvin) and Tyndall. The first resolution of this Congress reads as follows:[388]

1° On adoptera pour les mesures électriques les unités fondamentales: centimètre, masse du gramme, second (C. G. S.).

18.3 The Confusion Created by the so-called "Gaussian" System of Units

In modern times the most common systems of units are the International System of Units MKSA and the so-called "Gaussian" System of Units, which is based on CGS units. This so-called "Gaussian" system of units appears in many modern textbooks on electromagnetism, always associated with the basic CGS units of length, mass and time given by the centimeter, gram and second.[389] The name "Gaussian" system of units seems to be due to Helmholtz and

[387] [Tho74].
[388] [Con82, p. 42].
[389] [Jac75, p. 820, Table 4], [Gri89, p. 518, Table B.1], [HM95, p. 2] and [PM13, p. 764].

his student Hertz in the 1880's, when they utilized the expressions "Gauss units", "Gaussian units", "Gauss's system of units" or "absolute Gauss's measure".[390] Sometimes this system of units is also called "Gaussian-CGS units".

The electrostatic force between two charges in this "Gaussian" system is also given by Equation (18.5).

How many coulombs are there in 1 "Gaussian" unit of charge? The simplest way to answer this question is to consider two equal charges $q_1 = q_2 = q$ separated by 1 unit of distance in the "Gaussian" system, $r = 1\ cm = 10^{-2}\ m$, exerting on one another 1 dyne of force, $f = 1\ g \cdot cm/s^2 = 1\ dyne = 1 \times 10^{-5}\ kg \cdot m/s^2 = 1 \times 10^{-5}\ N$. Equation (18.8) then yields for this case:

$$q = 3.3 \times 10^{-10}\ C\ . \tag{18.11}$$

That is, a charge of $3.3 \times 10^{-10}\ C$ is equivalent to 1 "Gaussian" unit of charge $q = 1\ cm^{3/2} g^{1/2} s^{-1}$.

Therefore, while one unit of charge in the absolute system of units of Gauss and Weber corresponds to $3.3 \times 10^{-13}\ C$, one unit of charge in the "Gaussian" system of units corresponds to $3.3 \times 10^{-10}\ C$.

There are two main aspects which should be emphasized here, discussed below under (I) and (II).

(I) Although Gauss had initiated the measurement of terrestrial magnetism and of bar magnetism utilizing only units of length, mass and time, this whole research project of absolute measures was carried out by Wilhelm Weber. Weber utilized these three mechanical dimensions to effectively measure electric charges, electric currents, electrical resistance and electromotive force. He published these absolute measurements in his celebrated series of eight major Memoirs on *Electrodynamic Measurements* (*Elektrodynamische Maassbestimmungen*).[391]

Maxwell, for instance, described Weber's contributions in his *Treatise on Electricity and Magnetism* with the following words:[392]

> The introduction, by W. Weber, of a system of absolute units for the measurement of electrical quantities is one of the most important steps in the progress of the science. Having already, in conjunction with Gauss, placed the measurement of magnetic quantities in the first rank of methods of precision, Weber proceeded in his *Electrodynamic Measurements*[393] not only to lay down sound principles for fixing the units to be employed, but to make determinations of particular electrical quantities in terms of these units, with a degree of accuracy previously unattempted. Both the electromagnetic and the electrostatic systems of units owe their development and practical applications to these researches.

Therefore, it would be more appropriate to utilize the joint names of Gauss and Weber whenever referring to the absolute system of units.

---

[390] [Hel82], [Her00, pp. 138 and 199] and [Car15, p. 18].

[391] [Web46], [Web52b], [Web52a], [KW57], [Web64], [Web71], [Web78] and [Web94b]; with English translations in [Web07], [Web21b], [Web21c], [KW21], [Web21a], [Web72], [Web21d] and [Web08], respectively.

[392] [Max54, Vol. 2, Article 545, pp. 193-194].

[393] Maxwell was referring to a set of Memoirs written by Weber under the general title of *Elektrodynamische Maassbestimmungen*. The title of the Sixth Memoir published in 1871, [Web71], received this translation as *Electrodynamic Measurements* when it was published in 1872, [Web72].

(II) The so-called "Gaussian" system of units is based on the CGS system of units. Therefore its basic units of length, mass and time are the centimeter, the gram and the second. This fact creates a lot of confusion as Gauss and Weber always considered the millimeter, the milligram and the second as the basic units of length, mass and time in their absolute system of units. For instance, while one unit of charge in the absolute system of Gauss and Weber is equivalent to $q = 3.3 \times 10^{-13}$ C, one unit of charge in the "Gaussian" system of units is equivalent to $q = 3.3 \times 10^{-10}$ C.

I present here three suggestions to avoid this confusion and to clarify the situation:

- The expression "absolute system of units" should be reserved for the system of measurements introduced by Gauss and Weber. In this system all electric and magnetic properties are reduced to measurements of length, mass and time. In particular, the basic units of length, mass and time must be the millimeter, milligram and second.
- The names *Gauss*, *Weber*, *Gaussian* or *Weberian* should not be utilized when referring to any system of measurements based on CGS units.
- Moreover, the joint names of Gauss and Weber should be utilized whenever referring to the absolute system of units.

Bibliography

[AC11] A. K. T. Assis and J. P. M. d. C. Chaib. *Eletrodinâmica de Ampère: Análise do Significado e da Evolução da Força de Ampère, Juntamente com a Tradução Comentada de Sua Principal Obra sobre Eletrodinâmica.* Editora da Unicamp, Campinas, 2011.

[AC15] A. K. T. Assis and J. P. M. C. Chaib. *Ampère's Electrodynamics — Analysis of the Meaning and Evolution of Ampère's Force between Current Elements, together with a Complete Translation of His Masterpiece: Theory of Electrodynamic Phenomena, Uniquely Deduced from Experience.* Apeiron, Montreal, 2015. Available at `www.ifi.unicamp.br/~assis`.

[AG95] A. K. T. Assis and P. Graneau. The reality of Newtonian forces of inertia. *Hadronic Journal*, 18:271–289, 1995.

[AG96] A. K. T. Assis and P. Graneau. Nonlocal forces of inertia in cosmology. *Foundations of Physics*, 26:271–283, 1996.

[AH07] A. K. T. Assis and J. A. Hernandes. *The Electric Force of a Current: Weber and the Surface Charges of Resistive Conductors Carrying Steady Currents.* Apeiron, Montreal, 2007. Available at `www.ifi.unicamp.br/~assis`.

[AH09] A. K. T. Assis and J. A. Hernandes. *A Força Elétrica de uma Corrente: Weber e as Cargas Superficiais de Condutores Resistivos com Correntes Constantes*, volume 73 of *Coleção Acadêmica.* Edusp and Edufal, São Paulo and Maceió, 2009.

[AH13] A. K. T. Assis and J. A. Hernandes. *Elektrischer Strom und Oberflächenladungen: was Wilhelm Weber schon vor mehr als 150 Jahre wußte.* Apeiron, Montreal, 2013. German translation by H. Härtel. Available at `www.ifi.unicamp.br/~assis`.

[Amp] A.-M. Ampère. Analyse des mémoires lus par M. Ampère a l'Académie des Sciences, dans les séances des 18 et 25 septembre, des 9 et 30 octobre 1820. 20 pages. Reprint of Annales Générales des Sciences Physiques, Vol. 6, pp. 238-257 (1820).

[Amp20a] A.-M. Ampère. Analyse des mémoires lus par M. Ampère a l'Académie des Sciences, dans les séances des 18 et 25 septembre, des 9 et 30 octobre 1820. *Annales Générales des Sciences Physiques*, 6:238–257, 1820.

[Amp20b] A.-M. Ampère. Suite du Mémoire sur l'Action mutuelle entre deux courans électriques, entre un courant électrique et un aimant ou le globe terrestre, et entre deux aimants. *Annales de Chimie et de Physique*, 15:170–218, 1820.

[Amp21] A.-M. Ampère. Suite de la Note sur un Appareil à l'aide duquel on peut vérifier toutes les propriétés des conducteurs de l'électricité voltaïque, découvertes par M. Ampère. *Annales de Chimie et de Physique*, 18:313–333, 1821.

[Amp23] A.-M. Ampère. Mémoire sur la théorie mathématique des phénomènes électrodynamiques uniquement déduite de l'expérience, dans lequel se trouvent réunis les Mémoires que M. Ampère a communiqués à l'Académie royale des Sciences, dans les séances des 4 et 26 décembre 1820, 10 juin 1822, 22 décembre 1823, 12 septembre et 21 novembre 1825. *Mémoires de l'Académie Royale des Sciences de l'Institut de France*, 6:175–387, 1823. Despite this date, the work was only published in 1827.

[Amp26] A.-M. Ampère. *Théorie des Phénomènes Électro-dynamiques, Uniquement Déduite de l'Expérience*. Méquignon-Marvis, Paris, 1826.

[Amp65] A.-M. Ampère. On the Mathematical Theory of Electrodynamic Phenomena, Experimentally Deduced. In R. A. R. Tricker, *Early Electrodynamics — The First Law of Circulation*, pages 155–200, New York, 1965. Pergamon. Partial English translation by O. M. Blunn of Ampère's work "Mémoire sur la théorie mathématique des phénomènes électro-dynamiques uniquement déduite de l'expérience", *Mémoires de l'Académie royale des Sciences de l'Institut de France*, Vol. 6, pp. 175-387 (1823), issued 1827.

[Amp69] A. M. Ampère. The solenoid. Circuits and magnetic shells. In W. F. Magie, editor, *A Source Book in Physics*, pages 456–460, New York, 1969. McGraw-Hill. Extracts from "Théorie des phénomènes électrodynamiques uniquement déduite de l'expérience," Paris, 1826.

[Amp12] A.-M. Ampère. Mathematical Theory of Electrodynamic Phenomena, Uniquely Derived from Experiments. Translated by M. D. Godfrey. Available at `https://archive.org/details/AmpereTheorieEn` and `https://sites.google.com/site/michaeldgodfrey/physics-information-and-communication`, 2012.

[Ano42] Anonymous. Arrangement of magnetic observations. *Magazine of Science, and School of Arts*, 4(30):234–235, 1842. Issue of October 22, 1842.

[Ano89] Anonymous. Gauss and the electric telegraph. *Popular Science Monthly*, 34:396–399, 1889. Available at `https://en.wikisource.org/wiki/Popular_Science_Monthly/Volume_34/January_1889/Gauss_and_the_Electric_Telegraph`.

[Ano91] Anonymous. Obituary. Professor Wilhelm Weber. *The Electrician*, 27:245–246, 1891.

[Ano92] Anonymous. Sketch of William Edward Weber. *Popular Science Monthly*, 40:548–553, 1892. Available at `https://en.wikisource.org/wiki/Popular_Science_Monthly/Volume_40/February_1892/Sketch_of_William_Edward_Weber`.

[Arc89] T. Archibald. Energy and the mathematization of electrodynamics in Germany, 1845-1875. *Archives Internationales d'Histoire des Sciences*, 39:276–308, 1989.

[Arp87] H. Arp. *Quasars, Redhsifts and Controversies.* Interstellar Media, Berkeley, 1987.

[Arp98] H. Arp. *Seeing Red — Redshifts, Cosmology and Academic Science.* Apeiron, Montreal, 1998.

[Arp01] H. Arp. *O Universo Vermelho: Desvios para o Vermelho, Cosmologia e Ciência Acadêmica.* Editora Perspectiva, São Paulo, 2001. Portuguese translation by A. K. T. Assis and D. S. L. Soares.

[ARW02] A. K. T. Assis, K. Reich, and K. H. Wiederkehr. Gauss and Weber's creation of the absolute system of units in physics. *21st Century Science & Technology*, Vol. 15, No. 3:40–48, 2002.

[ARW04] A. K. T. Assis, K. Reich, and K. H. Wiederkehr. On the electromagnetic and electrostatic units of current and the meaning of the absolute system of units — For the 200th anniversary of Wilhelm Weber's birth. *Sudhoffs Archiv*, 88:10–31, 2004.

[Ass89] A. K. T. Assis. On Mach's principle. *Foundations of Physics Letters*, 2:301–318, 1989.

[Ass92] A. K. T. Assis. *Curso de Eletrodinâmica de Weber.* Setor de Publicações do Instituto de Física da Universidade Estadual de Campinas — UNICAMP, Campinas, 1992. Notas de Física IFGW Número 5. Available at `www.ifi.unicamp.br/~assis` and `www.bibliotecadigital.unicamp.br/document/?down=60362`.

[Ass94] A. K. T. Assis. *Weber's Electrodynamics.* Kluwer, Dordrecht, 1994. Nowadays this book is available through Springer. Doi: 10.1007/978-94-017-3670-1.

[Ass95] A. K. T. Assis. *Eletrodinâmica de Weber — Teoria, Aplicações e Exercícios.* Editora da Universidade Estadual de Campinas — UNICAMP, Campinas, 1st edition, 1995.

[Ass98] A. K. T. Assis. *Mecânica Relacional.* Editora do Centro de Lógica, Epistemologia e História da Ciência da UNICAMP/FAPESP, Campinas, 1998. Available at `www.ifi.unicamp.br/~assis`.

[Ass99a] A. K. T. Assis. *Relational Mechanics.* Apeiron, Montreal, 1999. Available at `www.ifi.unicamp.br/~assis`.

[Ass99b] A. K. T. Assis. *Uma Nova Física.* Editora Perspectiva, São Paulo, 1999.

[Ass03a] A. K. T. Assis. On the first electromagnetic measurement of the velocity of light by Wilhelm Weber and Rudolf Kohlrausch. In F. Bevilacqua and E. A. Giannetto, editors, *Volta and the History of Electricity*, pages 267–286. Università degli Studi di Pavia and Editore Ulrico Hoepli, Milano, 2003.

[Ass03b] A. K. T. Assis. Tradução de uma obra de Gauss. *Revista Brasileira de Ensino de Física*, 25:226–249, 2003. Paper in Portuguese: "Translation of a work by Gauss".

[Ass10] A. K. T. Assis. Wilhelm Weber's works translated into English. *21st Century Science & Technology*, Vol. 22, No. 4:67–69, 2010.

[Ass13] A. K. T. Assis. *Mecânica Relacional e Implementação do Princípio de Mach com a Força de Weber Gravitacional.* Apeiron, Montreal, 2013. Available at `www.ifi.unicamp.br/~assis`.

[Ass14a] A. K. T. Assis. *Relational Mechanics and Implementation of Mach's Principle with Weber's Gravitational Force.* Apeiron, Montreal, 2014. Available at `www.ifi.unicamp.br/~assis`.

[Ass14b] A. K. T. Assis. Text in memory of Peter Graneau (1921-2014). *Infinite Energy*, 114:14–15, 2014.

[Ass15] A. K. T. Assis. *Eletrodinâmica de Weber: Teoria, Aplicações e Exercícios.* Editora da Unicamp, Campinas, 2nd edition, 2015.

[AT17] A. K. T. Assis and M. Tajmar. Superconductivity with Weber's electrodynamics: the London moment and the Meissner effect. *Annales de la Fondation Louis de Broglie*, 42:307–350, 2017.

[AT19] A. K. T. Assis and M. Tajmar. Rotation of a superconductor due to electromagnetic induction using Weber's electrodynamics. *Annales de la Fondation Louis de Broglie*, 44:111–123, 2019.

[Ath89] W. A. Atherton. Gauss and Weber: An unlikely partnership. *Electronics and Wireless World*, 95:521–522, 1989.

[AW03] A. K. T. Assis and K. H. Wiederkehr. Weber quoting Maxwell. *Mitteilungen der Gauss-Gesellschaft*, 40:53–74, 2003.

[AWW11] A. K. T. Assis, K. H. Wiederkehr, and G. Wolfschmidt. *Weber's Planetary Model of the Atom*, volume 19 of *Nuncius Hamburgensis — Beiträge zur Geschichte der Naturwissenschaften.* Tredition Science, Hamburg, 2011. Edited by G. Wolfschmidt.

[AWW14] A. K. T. Assis, K. H. Wiederkehr, and G. Wolfschmidt. *O Modelo Planetário de Weber para o Átomo.* Apeiron, Montreal, 2014. Available at `www.ifi.unicamp.br/~assis`.

[AWW18] A. K. T. Assis, K. H. Wiederkehr, and G. Wolfschmidt. *Webers Planeten-Modell des Atoms.* Apeiron, Montreal, 2018. Translated by H. Härtel. Available at `www.ifi.unicamp.br/~assis`.

[BA98] M. Bueno and A. K. T. Assis. *Cálculo de Indutância e de Força em Circuitos Elétricos.* Editora da UFSC and Editora da UEM, Florianópolis and Maringá, 1st edition, 1998.

[BA01] M. d. A. Bueno and A. K. T. Assis. *Inductance and Force Calculations in Electrical Circuits*. Nova Science Publishers, Huntington, New York, 2001.

[BA15] M. Bueno and A. K. T. Assis. *Cálculo de Indutância e de Força em Circuitos Elétricos*. Apeiron, Montreal, 2nd edition, 2015. Available at `www.ifi.unicamp.br/~assis`.

[Bec29] Becquerel. Mémoire sur l'électro-chimie et l'emploi de l'électricité pour opérer des combinaisons. *Annales de Chimie et de Physique*, 41:5–45, 1829.

[Beu97] G. Beuermann. Greetings from Göttingen on Gauss-Weber article. *21st Century Science & Technology*, 10(1):5, 1997.

[Beu d] G. Beuermann, s. d. Die Sammlung historischwer physikalischer Apparate. Available at `https://www.uni-goettingen.de/en/47114.html`.

[Bev83] F. Bevilacqua. *The Principle of Conservation of Energy and the History of Classical Electromagnetic Theory*. La Goliardica Pavese, Pavia, 1983.

[Bev84] F. Bevilacqua. History of electrodynamics: W. Weber and R. Clausius on the principle of conservation of energy. In Rend. Accad. Naz. Sci. XL Mem. Sci. Fis. Natur., editor, *Proceedings of the Fifth National Congress on the History of Physics*, pages 217–225, Rome, 1984. Available at `https://media.accademiaxl.it/memorie/S5-VIX-P1-2-1985/Bevilacqua217-225.pdf`.

[Bev93] F. Bevilacqua. Helmholtz's Ueber die Erhaltung der Kraft: the emergence of a theoretical physicist. In D. Cahan, editor, *Hermann von Helmholtz and the Foundations of Nineteenth-Century Science*, pages 291–333, Berkeley, 1993. University of California Press.

[Bev94] F. Bevilacqua. Theoretical and mathematical interpretations of energy conservation: the Helmholtz-Clausius debate on central forces 1852-5. In L. Krüger, editor, *Universalgenie Helmholtz: Rückblick nach 100 Jahren*, pages 89–106, Berlin, 1994. Akademie Verlag.

[Bev95] F. Bevilacqua. The emergence of theoretical physics in the second half of the 19th century. In R. Zwilling, editor, *Natural Sciences and Human Thought*, pages 13–36, Berlin, 1995. Springer.

[Bev14] F. Bevilacqua. Energy: learning from the past. *Science & Education*, 23:1231–1243, 2014. Doi: 10.1007/s11191-014-9690-1.

[Bie71] K. -R. Biermann. Der Brief Alexander von Humboldts an Wilhelm Weber von Ende 1831 — ein bedeutendes Dokument zur Geschichte der Erforschung des Geomagnetismus. *Monatsberichte der Deutschen Akademie der Wissenschaften zu Berlin (DDR)*, 13, Heft 3:234–242, 1971.

[Blo05] C. Blondel, 2005. Ampère et l'Histoire de l'Électricité. Site of the CNRS coordinated by Christine Blondel. Available at `www.ampere.cnrs.fr`.

[Buc84] J. Z. Buchwald. *The Creation of Scientific Effects: Heinrich Hertz and Electric Waves*. The University of Chicago Press, Chicago, 1984.

[Buc85] J. Z. Buchwald. *From Maxwell to Microphysics — Aspects of Electromagnetic Theory in the Last Quarter of the Nineteenth Century.* The University of Chicago Press, Chicago, 1985.

[Buc20] O. M. Bucci. Electromagnetism without fields: from Ørsted through Ampère to Weber. *IEEE Antennas & Propagation Magazine*, 62:128–137, 2020. Doi: 10.1109/MAP.2020.2998847.

[BW21a] C. Blondel and B. Wolff. Ampère's force law: an obsolete formula? Translated by A. Butrica. Available at www.ampere.cnrs.fr/parcourspedagogique, 2021.

[BW21b] C. Blondel and B. Wolff. La force d'Ampère, une formule obsolète? Available at www.ampere.cnrs.fr/parcourspedagogique, 2021.

[CA09] J. P. M. d. C. Chaib and A. K. T. Assis. Sobre os efeitos das correntes elétricas (segunda parte) — Tradução da primeira obra de Ampère sobre eletrodinâmica. *Revista Brasileira de História da Ciência*, 2:118–145, 2009.

[Cah18] D. Cahan. *Helmholtz: A Life in Science.* The University of Chicago Press, Chicago, 2018.

[Can78] K. L. Caneva. From galvanism to electrodynamics: the transformation of German physics and its social context. *Historical Studies in the Physical Sciences*, 9:63–159, 1978.

[Car15] N. Carron, 2015. Babel of units. The evolution of units systems in classical electromagnetism. Available at https://arxiv.org/abs/1506.01951.

[Cha09] J. P. M. d. C. Chaib. *Análise do Significado e da Evolução do Conceito de Força de Ampère, juntamente com a Tradução Comentada de sua Principal Obra sobre Eletrodinâmica.* PhD thesis, University of Campinas — UNICAMP, Campinas, Brazil, 2009. Supervisor: A. K. T. Assis. Available at www.ifi.unicamp.br/~assis and http://repositorio.unicamp.br/jspui/handle/REPOSIP/277326.

[Cla76] P. Clark. Elkana on Helmholtz and the conservation of energy. *The British Journal for the Philosophy of Science*, 27:165–176, 1976.

[Con82] Congrès International des Électriciens, 1882. G. Masson. Paris.

[Cou88] C. A. Coulomb. Second mémoire sur l'électricité et le magnétisme. Où l'on détermine, suivant quelles loix le fluide magnétique, ainsi que le fluide électrique, agissent, soit par répulsion, soit par attraction. *Mémoires de l'Académie royale des Sciences. Année 1785*, 88:578–611, 1788. Published in 1788. Reprinted in A. Potier (ed.), Collection de Mémoires relatifs a la Physique, Volume 1: *Mémoires de Coulomb*, pp. 116-146 (Gauthiers-Villars, Paris, 1884).

[Cou90] C. A. Coulomb. Zweite Abhandlung über die Elektricität und den Magnetismus. In W. König, editor, *Vier Abhandlungen über die Elektricität und den Magnetismus von Coulomb*, pages 12–42. Wilhelm Engelmann, Leipzig, 1890. Uebersetzt von W. König.

[Cou35] A. Coulomb. Second memoir on electricity and magnetism. In W. F. Magie, editor, *A Source Book in Physics*, pages 413–420, New York, 1935. McGraw-Hill. Original publication in French in 1788.

[D'A81] S. D'Agostino. Weber and Maxwell on the discovery of the velocity of light in nineteenth century electrodynamics. In M. D. Grmek, R. S. Cohen, and G. Cimino, editors, *On Scientific Discovery*, pages 281–294, Dordrecht, 1981. Springer. Boston Studies in the Philosophy of Science, Vol. 34. Translated by J. A. Boffa. Doi: 10.1007/978-94-010-1284-3_14.

[D'A96] S. D'Agostino. Absolute systems of units and dimensions of physical quantities: a link between Weber's electrodynamics and Maxwell's electromagnetic theory of light. *Physis*, 33:5–51, 1996.

[D'A00] S. D'Agostino. *A History of the Ideas of Theoretical Physics: Essays on the Nineteenth and Twentieth Century Physics.* Kluwer, Dordrecht, 2000. Volume 213 of the Boston Studies in the Philosophy of Science.

[Dar93a] O. Darrigol. The electrodynamic revolution in Germany as documented by early German expositions. *Archive for the History of Exact Sciences*, 45:190–280, 1993.

[Dar93b] O. Darrigol. The electrodynamics of moving bodies from Faraday to Hertz. *Centaurus*, 36:245–260, 1993.

[Dar96] O. Darrigol. The electrodynamic origin of relativity theory. *Historical Studies in the Physical and Biological Sciences*, 26:241–312, 1996.

[Dar00] O. Darrigol. *Electrodynamics from Ampère to Einstein.* Oxford University Press, Oxford, 2000.

[DH30] H. W. Dove and A. v. Humboldt. Correspondirende Beobachtungen über die regelmäßsigen stündlichen Veränderungen und über die Perturbationen der magnetischen Abweichung im mittleren und östlichen Europa; gesammelt und verglichen von H. W. Dove, mit einem Vorwort von Alexander von Humboldt. *Annalen der Physik und Chemie*, 19:357–391, 1830.

[dlR29] A. de la Rive. Untersuchungen über die Ursachen der sogenannten Contactelektricität. *Annalen der Physik und Chemie*, 15:98–121, 1829.

[Dor07] N. E. Dorsey. Biographical sketch of Wilhelm Eduard Weber. *Terrestrial Magnetism and Atmospheric Electricity*, 12:125–128, 1907. Doi: 10.1029/TE012i003p00125.

[F.91] G. C. F. Wilhelm Eduard Weber. *Nature*, 44:229–230, 1891. Doi: 10.1038/044229b0. The author of this obituary was George Carey Foster (1835-1919).

[Far32a] M. Faraday. Experimental researches in electricity. *Philosophical Transactions*, 122:125–162, 1832. Read November 24, 1831. Reprinted in *Great Books of the Western World*, R. M. Hutchins (editor), (Encyclopaedia Britannica, Chicago, 1952), Vol. 45: Lavoisier, Fourier, Faraday. Pp. 265-285, §1-139.

[Far32b] M. Faraday. Experimental-Untersuchungen über Elektricität. *Annalen der Physik und Chemie*, 25:91–142, 1832.

[Far33] M. Faraday. Experimental researches in electricity. Fifth series. *Philosophical Transactions*, 123:675–710, 1833. Read June 20, 1833. Reprinted in *Great Books of the Western World*, R. M. Hutchins (editor), (Encyclopaedia Britannica, Chicago, 1952), Vol. 45: Lavoisier, Fourier, Faraday. Pp. 327-346, §450-563.

[Far34a] M. Faraday. Experimental researches in electricity. Seventh series. *Philosophical Transactions*, 124:77–122, 1834. Read January 23, February 6 and 13, 1834. Reprinted in *Great Books of the Western World*, R. M. Hutchins (editor), (Encyclopaedia Britannica, Chicago, 1952), Vol. 45: Lavoisier, Fourier, Faraday. Pp. 361-390, §661-874.

[Far34b] M. Faraday. Von der elektro-chemischen Zersetzung — Fortsetzung. *Annalen der Physik und Chemie*, 33:301–331 and 433–451, 1834.

[Far34c] M. Faraday. Von der elektrochemischen Zersetzung. *Annalen der Physik und Chemie*, 32:401–453, 1834.

[Far89] M. Faraday. *Experimental-Untersuchungen über Elektricität*, volume I. Springer, Berlin, 1889. Deutsche Uebersetzung von S. Kalischer.

[Far11] M. Faraday. Pesquisas experimentais em eletricidade. *Caderno Brasileiro de Ensino de Física*, 28:152–204, 2011. Portuguese translation by A. K. T. Assis and L. F. Haruna. Doi: 10.5007/2175-7941.2011v28n1p152.

[Fec32] G. T. Fechner. *Repertorium der Experimentalphysik: enthaltend eine vollständige Zusammenstellung der neuern Fortschritte dieser Wissenschaft; Als Supplement zu neuern Lehr- und Wörterbüchern der Physik*, volume 1. Leopold Voss, Leipzig, 1832.

[Fer07] J. Ferreirós. The rise of pure mathematics as arithmetic with Gauss. In C. Goldstein, N. Schappacher, and J. Schwermer, editors, *The Shaping of Arithmetic after C. F. Gauss's Disquisitiones Arithmeticae*, pages 207–240. Springer, Berlin, 2007.

[Fer20] J. Ferreirós. Gauss and the mathematical background to standardisation. *Journal of History of Science and Technology*, 14:32–51, 2020.

[Fey33a] E. Feyerabend. Der erste praktisch verwendete elektrische Telegraph. *Technikgeschichte*, 22:83–87, 1933.

[Fey33b] E. Feyerabend. *Der Telegraph von Gauss und Weber im Werden der elektrischen Telegraphie*. Reichsportminsterium, Berlin, 1933.

[Fra81] O. I. Franksen. *H. C. Ørsted — A Man of the Two Cultures*. Strandbergs Forlag, Birkerød, 1981.

[Fri82] H. Fricke. *Two rival programmes in 19th. century classical electrodynamics: action-at-a-distance versus field theories.* PhD thesis, The London School of Economics and Political Science, London, 1982. Available at `http://etheses.lse.ac.uk/3430/`.

[GA94] P. Graneau and A. K. T. Assis. Kirchhoff on the motion of electricity in conductors. *Apeiron*, 19:19–25, 1994.

[Gal93] G. Galeczki. The ultimate speed and Weber's potential. *Physics Essays*, 6:448–450, 1993.

[Gar01] E. Garber. *The Language of Physics: The Calculus and the Development of Theoretical Physics in Europe, 1750-1914.* Springer Science, New York, 2001.

[Gaua] C. F. Gauss. Gauss-Gesellschaft e. V. [Gauss Society] Göttingen. Founded in May 17, 1962. Homepage: `http://www.gauss-gesellschaft-goettingen.de/gauss-e.htm`.

[Gaub] C. F. Gauss. The complete correspondence of Carl Friedrich Gauß. Site of the Department "Digital Library" of the State and University Library in Göttingen on behalf of the Academy of Sciences in Göttingen. The metadata of the letters have been made available by Prof. Dr. Menso Folkerts. Available at `https://gauss.adw-goe.de`.

[Gau32] C. F. Gauss. Anzeige der "Intensitas Vis Magneticae Terrestris ad Mensuram Absolutam Revocata". *Göttingische gelehrte Anzeigen*, 205-207:2041–2058, 1832. Vol. 205 from December 24, 1832, pp. 2041-2048; Vols. 206 and 207 from December 27, 1832, pp. 2049-2058. Reprinted in C. F. Gauss's *Werke*, Vol. 5, pp. 293-304 (Königlichen Gesellschaft der Wissenschaften, Göttingen, 1867). Corrected version in: Astronomische Nachrichten 10, 1833, Nr. 238, pp. 349-360: Anzeige der Abhandlung des Herrn Hofraths Gauß: Intensitas vis magneticae terrestris ad mensuram absolutam revocata.

[Gau33a] C. F. Gauss. [abstract of the paper:] intensitas vis magneticae terrestris ad mensuram absolutam revocata. *Philosophical Magazine*, 2:291–299, 1833.

[Gau33b] C. F. Gauss. Die Intensität der erdmagnetischen Kraft, zurückgeführt auf absolutes Maass. *Annalen der Physik und Chemie*, 28:241–273 and 591–615, 1833. Translated by J. C. Poggendorff.

[Gau34a] C. F. Gauss. 128. Stuck. Den 9. August 1834. *Göttingische gelehrte Anzeigen*, 2:1265–1274, 1834. Reprinted in C. F. Gauss's *Werke*, Vol. 5, pp. 519-525 (Königlichen Gesellschaft der Wissenschaften, Göttingen, 1867).

[Gau34b] C. F. Gauss. Beobachtungen der magnetischen Variation in Göttingen und Leipzig, am 1. und 2. October 1834. *Annalen der Physik und Chemie*, 33:426–433, 1834. Reprinted in C. F. Gauss's *Werke*, Vol. 5, pp. 525-528 (Königlichen Gesellschaft der Wissenschaften, Göttingen, 1867).

[Gau34c] C. F. Gauss. Mesure absolue de l'Intensité du Magnétisme terrestre. *Annales de Chimie et de Physique*, 57:5–69, 1834.

[Gau35a] C. F. Gauss. Bericht von neuerlich in Göttingen angestellten magnetischen Beobachtungen. *Annalen der Physik und Chemie*, 34:546–556, 1835.

[Gau35b] C. F. Gauss. Berichte über die in dem magnetischen Observatorium, und in Verbindung damit anderwärts gemachten Beobachtungen. *Göttingische gelehrte Anzeigen*, 1:345–357, 1835. 36. Stuck der *Göttingische gelehrte Anzeigen*, den 7. Merz 1835. Reprinted in C. F. Gauss's *Werke*, Vol. 5, pp. 528-536 (Königlichen Gesellschaft der Wissenschaften, Göttingen, 1867).

[Gau36a] C. F. Gauss, 1836. Ob isměrenii zemnago magnitizma. (Soč[inenie] Karl[a] Frid[richa] Gaussa). Per[evël] A. Drašusov. Učenyja zapiski Imperatorskago universiteta čast 11, 1836, Nr. 7 (Januar), str. 3-22; Nr. 8 (Februar), str. 246-271; Nr. 9 (März), str. 341-381.

[Gau36b] C. F. Gauss. Erdmagnetismus und Magnetometer. In H. C. Schumacher, editor, *Jahrbuch für 1836*, pages 1–47. J. G. Cotta'schen, Stuttgart, 1836. Reprinted in C. F. Gauss's *Werke*, Vol. 5, pp. 315-344 (Königlichen Gesellschaft der Wissenschaften, Göttingen, 1867).

[Gau37a] C. F. Gauss. [abstract of the paper:] intensitas vis magneticae terrestris ad mensuram absolutam revocata. *Abstracts of the Papers Printed in the Philosophical Transactions of the Royal Society of London, from 1830 to 1837 inclusive*, 3:166–174, 1837.

[Gau37b] C. F. Gauss. Einleitung. In C. F. Gauss and W. Weber, editors, *Resultate aus den Beobachtungen des magnetischen Vereins im Jahre 1836*, volume I, pages 3–12. Dieterichschen Buchhandlung, Göttingen, 1837. Reprinted in C. F. Gauss's *Werke*, Vol. 5, pp. 345-351 (Königlichen Gesellschaft der Wissenschaften, Göttingen, 1867).

[Gau37c] C. F. Gauss. Misura assoluta dell' intensità della forza magnetica terrestre. *Effemeridi astronomiche di Milano*, primo supplemento:3–132, 1837. Tradotta e commentata da P. Frisiani.

[Gau38a] C. F. Gauss. Sur le magnétisme terrestre et sur l'association pour les observations magnétiques — Préface. In A. Quetelet, editor, *Correspondance Mathématique et Physique*, pages 41–49. Société Belge de Librairie, Bruxelles, 1838. Tome X et Tome Second de la Troisième Série.

[Gau38b] C. F. Gauss. Ueber ein neues, zunächst zur unmittelbaren Beobachtung der Veränderungen in der Intensität des horizontalen Theils des Erdmagnetismus bestimmtes Instrument. In C. F. Gauss and W. Weber, editors, *Resultate aus den Beobachtungen des magnetischen Vereins im Jahre 1837*, volume II, chapter I, pages 1–19. Dieterichschen Buchhandlung, Göttingen, 1838. Reprinted in Carl Friedrich Gauss *Werke*, Vol. 5, pp. 357-373 (Königlichen Gesellschaft der Wissenschaften, Göttingen, 1867).

[Gau40] C. F. Gauss. Allgemeine Lehrsätze in Beziehung auf die im Verkehrten Verhältnisse des Quadrats der Entfernung wirkenden Anziehungs- und Abstossungs-kräfte. In C. F. Gauss and W. Weber, editors, *Resultate aus den*

Beobachtungen des magnetischen Vereins im Jahre 1839, volume IV, chapter I, pages 1–51. Weidmannschen Buchhandlung, Leipzig, 1840. Reprinted in Carl Friedrich Gauss *Werke*, Vol. 5, pp. 195-242 (Königlichen Gesellschaft der Wissenschaften, Göttingen, 1867).

[Gau41a] C. F. Gauss. Intensitas vis magneticae terrestris ad mensuram absolutam revocata. *Commentationes Societatis Regiae Scientiarum Goettingensis Recentiores*, 8:3–44, 1841. Delivered before the Society in 15 December 1832. Reprinted in C. F. Gauss's *Werke*, Vol. 5, pp. 79-118 (Königliche Gesellschaft der Wissenschaften, Göttingen, 1867).

[Gau41b] C. F. Gauss. Introduction. In R. Taylor, editor, *Scientific Memoirs,* Vol. 2, pages 20–25, London, 1841. Richard and John E. Taylor. Translated by Mr. W. Francis. The translation has been revised by Professor Lloyd and Major Sabine.

[Gau41c] C. F. Gauss. On a new instrument for the direct observation of the changes in the intensity of the horizontal portion of the terrestrial magnetic force. In R. Taylor, editor, *Scientific Memoirs,* Vol. 2, pages 252–267, London, 1841. Richard and John E. Taylor. From the *Resultate aus den Beobachtungen des magnetischen Vereins im Jahre 1837*, Volume II, Chapter I, pp. 1-19.—Herausgegeben von Carl Friederich Gauss und Wilhelm Weber. Göttingen, 1838. Translated by Mr. William Francis, and revised by Major Sabine and Professor Lloyd.

[Gau41d] C. F. Gauss. Vorschriften zur Berechnung der magnetischen Wirkung, welche ein Magnetstab in der Ferne ausübt. In *Resultate aus den Beobachtungen des magnetischen Vereins im Jahre 1840*, volume V, chapter II, pages 26–34. Weidmannschen Buchhandlung, Leipzig, 1841. Reprinted in Carl Friedrich Gauss *Werke*, Vol. 5, pp. 427-435 (Königlichen Gesellschaft der Wissenschaften, Göttingen, 1867).

[Gau43] C. F. Gauss. General propositions relating to attractive and repulsive forces acting in the inverse ratio of the square of the distance. In R. Taylor, editor, *Scientific Memoirs,* Vol. 3, pages 153–196, London, 1843. Richard and John E. Taylor.

[Gau45] C. F. Gauss, 1845. Zur Elektrodynamik. Letter [20] from C. F. Gauss to W. Weber from March 19, 1845. Reprinted in Carl Friedrich Gauss *Werke*, Vol. 5, pp. 627-629 (Königlichen Gesellschaft der Wissenschaften, Göttingen, 1867).

[Gau94] C. F. Gauss. Die Intensität der erdmagnetischen Kraft auf absolutes Maass zurückgeführt. In E. Dorn, editor, *Ostwald's Klassiker der exakten Wissenschaften, Vol. 53.* Wilhelm Engelmann Verlag, Leipzig, 1894. Translation by A. Kiel, notes by E. Dorn. Reprinted in: *Carl Friedrich Gauss — Werke — Supplement* — Band 3: Varia: 15 Abhandlungen in deutscher Übersetzung. Mit einer Einleitung, einer Bibliographie und Registern herausgegeben von Karin Reich. Hrsg. von der Akademie der Wissenschaften zu Göttingen, Universitätsverlag Göttingen 2019, S. 533-594.

[Gau35] C. F. Gauss. The absolute measure of magnetic force. In W. F. Magie, editor, *A Source Book in Physics*, pages 519–524, New York, 1935. McGraw-Hill.

[Gau52] C. F. Gauss, 1952. Intensivnost semnoj magnitnoj sily, privedennaja k absoljutnoj mere. In: B. M. Janovskij (Hrsg): Karl Fridrich Gauss. Izbrannye trudy po zemnomu magnetizmu. Perevod A. N. Krylova. Moskau 1952, str. 23-75.

[Gau75] C. F. Gauss. Letter from Gauss to Schumacher, April 29, 1845. In *Carl Friedrisch Gauss, Werke, Ergänzungsreihe, Vol. V, Briefwechsel C. F. Gauss — H. C. Schumacher, Part 2*, pages 436–440. Georg Olms Verlag, Hildesheim, 1975.

[Gau03] C. F. Gauss. The intensity of the earth's magnetic force reduced to absolute measurement. Translated in 1995 from the German by Susan P. Johnson, edited by L. Hecht. Available since 2003 at `http://21sci-tech.com/translation.html`, 2003.

[GG93] P. Graneau and N. Graneau. *Newton Versus Einstein — How Matter Interacts with Matter.* Carlton Press, New York, 1993.

[GG96] P. Graneau and N. Graneau. *Newtonian Electrodynamics.* World Scientific, Singapore, 1996.

[Gib97] P. Gibbs, 1997. Why is c the symbol for the speed of light? Available at `http://math.ucr.edu/home/baez/physics/Relativity/SpeedOfLight/c.html` and `http://vixra.org/abs/1901.0004`.

[Gil71a] C. S. Gillmor. *Coulomb and the Evolution of Physics and Engineering in Eighteenth-Century France.* Princeton University Press, Princeton, 1971.

[Gil71b] C. S. Gillmor. Coulomb, Charles Augustin. In C. C. Gillispie, editor, *Dictionary of Scientific Biography,* Vol. 3, pages 439–447. Charles Scribner's Sons, New York, 1971.

[Glu99] A. G. Gluckman. On W. E. Weber's "Grundprincip der Elektrodynamik" and related earlier and contemporary studies. *Physics Essays*, 12:682–698, 1999.

[GM95] G. Galeczki and P. Marquardt. Über die longitudinalen Ampèreschen Kräfte und Webers Elektrodynamik. *Fusion*, 16:15–17, 1995.

[Gra94] P. Graneau. *Ampere-Neumann Electrodynamics of Metals.* Hadronic Press, Palm Harbor, 2nd edition, 1994.

[Gri89] D. J. Griffiths. *Introduction to Electrodynamics.* Prentice Hall, Englewood Cliffs, 2nd edition, 1989.

[Gro39] W. R. Grove. Volta'sche Säule von großser elektro-chemischer Kraft. *Annalen der Physik und Chemie*, 48:300–304, 1839.

[GT14] K. H. Glassmeier and B. T. Tsurutani. Carl Friedrich Gauss - General Theory of Terrestrial Magnetism - a revised translation of the German text. *History of Geo- and Space Sciences*, 5:11–62, 2014. Doi: 10.5194/hgss-5-11-2014.

[GW37] C. F. Gauss and W. Weber. *Resultate aus den Beobachtungen des magnetisches Vereins im Jahre 1836*, volume I. Dieterichschen Buchhandlung, Göttingen, 1837.

[GW38a] C. F. Gauss and W. Weber. *Resultate aus den Beobachtungen des magnetisches Vereins im Jahre 1837*, volume II. Dieterichschen Buchhandlung, Göttingen, 1838.

[GW38b] C. F. Gauss and W. Weber. Sur le magnétisme terrestre et sur l'association pour les observations magnétiques. In A. Quetelet, editor, *Correspondance Mathématique et Physique*, pages 41–90. Société Belge de Librairie, Bruxelles, 1838. Tome X et Tome Second de la Troisième Série.

[GW39a] Gauss and Weber. Remarks on the construction of magnetic observatories and the instruments which they should contain. *The Annals of Electricity, Magnetism, & Chemistry; and Guardian of Experimental Science*, 3:92–108, 1839.

[GW39b] C. F. Gauss and W. Weber. *Resultate aus den Beobachtungen des magnetisches Vereins im Jahre 1838*, volume III. Dieterichschen Buchhandlung, Göttingen, 1839.

[GW40a] C. F. Gauss and W. Weber. *Atlas des Erdmagnetismus nach den Elementen der Theorie Entworfen — Supplement zu den Resultaten aus den Beobachtungen des magnetisches Vereins unter Mitwirkung von C. W. B. Goldschmidt.* Weidmann'sche Buchhandlung, Leipzig, 1840. Reprinted in C. F. Gauss's *Werke*, Vol. 12, pp. 335-408 (Königlichen Gesellschaft der Wissenschaften, Göttingen, 1929).

[GW40b] C. F. Gauss and W. Weber. *Resultate aus den Beobachtungen des magnetisches Vereins im Jahre 1839*, volume IV. Weidmannschen Buchhandlung, Leipzig, 1840.

[GW41a] C. F. Gauss and W. Weber. *Resultate aus den Beobachtungen des magnetisches Vereins im Jahre 1840*, volume V. Weidmannschen Buchhandlung, Leipzig, 1841.

[GW41b] C. F. Gauss and W. Weber. Results of the observations made by the Magnetic Association in the year 1836. In R. Taylor, editor, *Scientific Memoirs*, Vol. 2, pages 20–97, London, 1841. Taylor and Francis. Translated by Mr. W. Francis. The translation has been revised by Professor Lloyd and Major Sabine. Available at `https://www.biodiversitylibrary.org/bibliography/2501#/summary`.

[GW43] C. F. Gauss and W. Weber. *Resultate aus den Beobachtungen des magnetisches Vereins im Jahre 1841*, volume VI. Weidmannschen Buchhandlung, Leipzig, 1843.

[GW66] C. F. Gauss and W. Weber. Results of the observations made by the Magnetic Association in the year 1836. In R. Taylor, editor, *Scientific Memoirs*, Vol. 2, pages 20–97, New York, 1966. Johnson Reprint Corporation. Translated by Mr. W. Francis. The translation has been revised by Professor Lloyd and Major Sabine.

[GW96] C. F. Gauss and W. E. Weber. Text of the Gauss-Weber correspondence. *21st Century Science & Technology*, 9(3):41–43, 1996. Letters transcribed from the German script by K. Krause and A. Hartmann. Edited by L. Hecht. English translation by Susan P. Johnson. Available at `https://`

21sci-tech.com/articles/Atomic_Science.pdf and https://archive.org/details/WeberAmpereElectrodynamicsHistory.

[Han19] C. Hansteen. *Untersuchungen über den Magnetismus der Erde*, volume I: Die mechanischen Erscheinungen des Magneten. Christiania, Kopenhagen, 1819. Translated by P. T. Hanson.

[Han25] C. Hansteen. Beobachtungen über die Intensität des Magnetismus im nördlichen Europa. *Annalen der Physik unc Chemie*, 79:225–270 and 353–428, 1825.

[Han27] C. Hansteen. Isodynamische Linien für die ganze magnetische Kraft. *Annalen der Physik und Chemie*, 85:49–66 and 229–244, 1827.

[Har82] P. M. Harman. *Energy, Force, and Matter — The Conceptual Development of Nineteenth-Century Physics*. Cambridge University Press, Cambridge, 1982.

[HB04] A. v. Humboldt and J. B. Biot. Sur les variations du magnétisme terrestre a différentes latitudes. *Journal de Physique, de Chimie, d'Histoire Naturelle et des Arts*, 59:429–450, 1804.

[HB05] A. v. Humboldt and J. B. Biot. Ueber die Variationen des Magnetismus der Erde in verschidenen Breite. *Annalen der Physik und Chemie*, 20:257–298, 1805.

[Hec88] L. Hecht. The geometric basis for the periodicity of the elements. *21st Century Science & Technology*, May-June:18–30, 1988. Available at https://21sci-tech.com/Articles%202004/Spring2004/Periodicity.pdf.

[Hec96a] L. Hecht. Experimental apparatus and instrumentation. *21st Century Science & Technology*, 9(3):35–40, 1996.

[Hec96b] L. Hecht. The significance of the 1845 Gauss-Weber correspondence. *21st Century Science & Technology*, 9(3):22–34, 1996. Available at https://21sci-tech.com/articles/Atomic_Science.pdf and https://archive.org/details/WeberAmpereElectrodynamicsHistory.

[Hec97] L. Hecht. Wilhelm Weber: giant of 19th century physics. *21st Century Science & Technology*, 10(2):53–54, 1997.

[Hec00] L. Hecht. Science: to be, or not to be — or, how I discovered the swindle of special relativity. *21st Century Science & Technology*, 12(4):2–4 and 7, 1999-2000.

[Hec01] L. Hecht. Should the law of gravity be repealed? The suppressed electrodynamics of Ampère-Gauss-Weber. *21st Century Science & Technology*, 14(1):2–7, 2001.

[Hec04] L. Hecht. The life and work of Dr. Robert J. Moon. *21st Century Science & Technology*, Fall:8–20, 2004. Available at https://21sci-tech.com/Articles%202005/moon_F04.pdf.

[Hec07a] L. Hecht. The Ampère angular force and the Newton hoax. *21st Century Science & Technology*, 20(1-2):53–55 and 71, 2007. This article also appeared in the Executive Intelligence Review, vol. 34, No. 15, April 13, 2007. Available at http://21sci-tech.com/Subscriptions/Spring%202007%20ONLINE/Spring-Summer_2007.pdf.

[Hec07b] L. Hecht. Wilhelm Weber defended. *21st Century Science & Technology*, 20(3):72, 2007. Available at `https://21sci-tech.com/Subscriptions/Fall%202007%20Online/Books.pdf`.

[Hee11] P. Heering. Tools for investigation, tools for instruction: potential transformations of instruments in the transfer from research to teaching. In P. Heering and R. Wittje, editors, *Learning by Doing, Experiments and Instruments in the History of Science Teaching*, pages 15–30, Stuttgart, 2011. Franz Steiner.

[Hei82] J. L. Heilbron. *Elements of Early Modern Physics*. University of California Press, Berkeley, 1982.

[Hei99] J. L. Heilbron. *Electricity in the 17th and 18th Centuries — A Study in Early Modern Physics*. Dover, New York, 1999.

[Hel82] H. Helmholtz. Ueber absolute Maasssysteme für electrische und magnetische Grössen. *Annalen der Physik und Chemie*, 17:42–54, 1882.

[Hen04] K. Hentschel. Leben und Werke des Instrumentmachers Moritz Meyerstein (1808-82). *Einbecker Jahrbuch*, 49:157–184, 2004.

[Hen05] K. Hentschel. *Gaußens unsichtbare Hand: Der Universitätsmechanicus und Maschineninspector Moritz Meyerstein — Ein Instrumentbauer im 19. Jahrhundert*. Abhandlungen der Göttinger Akademie der Wissenschaften, Göttingen, 2005. Mathem.-Physik. Klasse, Reihe 3, Nr. 52.

[Hen07] K. Hentschel. Gauß, Meyerstein and Hanoverian metrology. *Annals of Science*, 64:41–72, 2007.

[Hen20] K. Hentschel. The 'invisible hand' of Carl Friedrich Gauß— retracing the life of Moritz Meyerstein, a 19th century instrument maker and Universitäts-Mechanicus. In C. Forstner and M. Walker, editors, *Biographies in the History of Physics: Actors, Objects, Institutions*, chapter 2, pages 13–36. Springer, Berlin, 2020. Translated by A. M. Hentschel.

[Her00] H. Hertz. *Electric Waves*. Dover, New York, 2nd edition, 1900.

[Hes55] M. B. Hesse. Action at a distance in classical physics. *Isis*, 46:337–353, 1955.

[Hes61] M. B. Hesse. *Forces and Fields: A Study of Action at a Distance in the History of Physics*. Thomas Nelson and Sons, London, 1961.

[HM95] M. A. Heald and J. B. Marion. *Classical Electromagnetic Radiation*. Saunders, Fort Worth, 3rd edition, 1995.

[HS04] L. Hecht and C. B. Stevens. New explorations with the Moon model. *21st Century Science & Technology*, Fall:58–73, 2004. Available at `http://21sci-tech.com/Articles%202005/MoonModel_F04.pdf`.

[Hum01] A. v. Humboldt. Neue physikalische Beobachtungen im spanischen Amerika. Aus Briefen an Fourcroy und Lalande. *Annalen der Physik und Chemie*, 7:329–347, 1801.

[Hum29a] A. v. Humboldt. Beobachtungen der Intensität magnetischer Kräfte und der magnetischen Neigung, angestellt in den Jahren 1798 bis 1803, von 48° 50' N.B. bis 12° S.B. und 3° 2' O.L. bis 106° 22' W.L., in Frankreich, Spanien, den canarischen Inseln, dem atlantischen Ocean, America und der Südsee. *Annalen der Physik und Chemie*, 15:336–355, 1829.

[Hum29b] A. v. Humboldt. Ueber die Mittel, die Ergründung einiger Phänomene des tellurischen Magnetismus zu erleichtern. *Annalen der Physik und Chemie*, 15:319–336, 1829.

[Hun21] B. J. Hunt. *Imperial Science: Cable Telegraphy and Electrical Physics in the Victorian British Empire.* Cambridge University Press, Cambridge, 2021.

[Jac39] M. Jacobi. Ueber das chemische und das magnetische Galvanometer. *Annlen der Physik und Chemie*, 48:26–57, 1839. Mitgetheilt vom Hen. Verfasser aus den Bullet. scient. der Petersburger Academie, T. V.

[Jac75] J. D. Jackson. *Classical Electrodynamics.* John Wiley & Sons, New York, second edition, 1975.

[Jac06] M. W. Jackson. *Harmonious Triads: Physicists, Musicians, and Instrument Makers in Nineteenth-Century Germany.* The MIT Press, Cambridge, 2006.

[JM86] C. Jungnickel and R. McCormmach. *Intellectual Mastery of Nature — Theoretical Physics from Ohm to Einstein*, volume 1-2. University of Chicago Press, Chicago, 1986.

[JM17] C. Jungnickel and R. McCormmach. *The Second Physicist: On the History of Theoretical Physics in Germany.* Springer, Cham, 2017.

[Joh97] L. Johnson. In memoriam: Susan P. Johnson. *Executive Intelligence Review*, 24:58, 1997.

[Jou41] J. P. Joule. On the heat evolved by metallic conductors of electricity, and in the cells of a battery during electrolysis. *Philosophical Magazine*, 19:260–277, 1841.

[Kär02] M. Kärn. Das erdmagnetische Observatorium in der Scheune: Messungen mit dem originalgetreuen Nachbau eines Magnetometers vom Gauß und Weber. *Mitteilungen der Gauss-Gesellschaft*, 39:23–52, 2002.

[Kir49] G. Kirchhoff. Ueber eine Ableitung der Ohm'schen Gesetze, welche sich an die Theorie der Elektrostatik anschliesst. *Annalen der Physik und Chemie*, 78:506–513, 1849. Reprinted in G. Kirchhoff's *Gesammelte Abhandlungen* (Barth, Leipzig, 1882), pp. 49-55.

[Kir50] G. Kirchhoff. On a deduction of Ohm's law, in connexion with the theory of electrostatics. *Philosophical Magazine*, 37:463–468, 1850.

[Kir57a] G. Kirchhoff. On the motion of electricity in wires. *Philosophical Magazine*, 13:393–412, 1857. Available at `https://archive.org/stream/londonedinburghp13maga#page/392/mode/2up`.

[Kir57b] G. Kirchhoff. Ueber die Bewegung der Elektricität in Drähten. *Annalen der Physik und Chemie*, 100:193–217, 1857. Reprinted in G. Kirchhoff's *Gesammelte Abhandlungen* (Barth, Leipzig, 1882), pp. 131-154.

[Kir57c] G. Kirchhoff. Ueber die Bewegung der Elektricität in Leitern. *Annalen der Physik und Chemie*, 102:529–544, 1857. Reprinted in G. Kirchhoff's *Gesammelte Abhandlungen* (Barth, Leipzig, 1882), pp. 154-168.

[Kir56] F. Kirchner. Über die Bestimmung der Lichtgeschwindigkeit aus elektromagnetischen Messungen nach W. Weber und R. Kohlrausch. *Die Naturwissenschaften*, 43:529–533, 1956.

[Kir57] F. Kirchner. Determination of the velocity of light from electromagnetic measurements according to W. Weber and R. Kohlrausch. *American Journal of Physics*, 25:623–629, 1957. Translated by A. Akeley.

[Koh83] F. Kohlrausch. *An Introduction to Physical Measurements — with appendices on absolute electrical measurement, etc.* J. & A. Churchill, London, 2nd edition, 1883. Translated from the fourth German edition by T. H. Waller and H. R. Procter.

[Koh10] F. Kohlrausch. *Lehrbuch der praktischen Physik.* B. G. Teubner, Leipzig, 1910. Elfte, stark vermehrte Auflage des Leitfadens der praktischen Physik.

[Kra10] H. Kragh, 2010. Before Bohr: Theories of atomic structure 1850-1913. RePoSS: Research Publications on Science Studies # 10. Aarhus: Centre for Science Studies, University of Aarhus, Denmark. Research group: History and philosophy of science.

[Kra12] H. Kragh. Zöllner's universe. *Physics in Perspective*, 14:392–420, 2012.

[KW57] R. Kohlrausch and W. Weber. Elektrodynamische Maassbestimmungen insbesondere Zurückführung der Stromintensitäts-Messungen auf mechanisches Maass. *Abhandlungen der Königlich Sächsischen Gesellschaft der Wissenschaften zu Leipzig, mathematisch-physischen Classe*, 3:221–292, 1857. Reprinted in Wilhelm Weber's *Werke*, Vol. 3, H. Weber (ed.), (Springer, Berlin, 1893), pp. 609-676.

[KW21] R. Kohlrausch and W. Weber, 2021. Electrodynamic measurements, especially attributing mechanical units to measures of current intensity. Third version posted in February 2021 at `www.ifi.unicamp.br/~assis`. Translated by D. H. Delphenich and edited by A. K. T. Assis.

[LA98] E. K. Lauridsen and N. Abrahamsen. The history of astatic magnet systems and suspensions. *Centaurus*, 40:135–169, 1998.

[LB67] D. Lardner and E. B. Bright. *The Electric Telegraph.* James Walton, London, 3rd edition, 1867. A new edition revised and re-written by E. B. Bright.

[Lin05] I. V. Lindell. Evolution of electromagnetics in the 19th century. *Advances in Radio Science*, 3:23–25, 2005.

[Lin06] I. V. Lindell. Evolution of electromagnetics in the 19th century. In T. K. Sarkar et al., editor, *History of Wireless*, pages 165–188, New York, 2006. Wiley.

[Mar90] M. Marshall. The theme of quantification and the hidden Weber in the early work of Gustav Theodor Fechner. *Canadian Psychology*, 31:45–53, 1990.

[Max73] J. C. Maxwell. *A Treatise on Electricity and Magnetism*, volume I. Clarendon Press, Oxford, 1873.

[Max54] J. C. Maxwell. *A Treatise on Electricity and Magnetism.* Dover, New York, 1954. Unabridged republication of the third edition of 1891.

[May72] T. Mayer. *The Unpublished Writings of Tobias Mayer*, volume III: The Theory of the Magnet and its Application to Terrestrial Magnetism. Vandenhoeck & Ruprecht, Göttingen, 1972. Edited and translated with introductions by E. G. Forbes.

[MB91] S. R. C. Malin and D. R. Barraclough. Humboldt and the Earth's magnetic field. *Quarterly Journal of the Royal Astronomical Society*, 32:279–293, 1991.

[Men06] K. S. Mendelson. The story of c. *American Journal of Physics*, 74:995–997, 2006. Doi: 10.1119/1.2238887.

[Men15] K. S. Mendelson. On an early proposal for a unified system of units. *American Journal of Physics*, 83:183–185, 2015. Doi: 10.1119/1.4896353.

[Mey90] J. Meya. *Elektrodynamik im 19. Jahrhundert: Rekonstruktion ihrer Entwicklung als Konzept einer redlichen Vermittlung.* Springer Fachmedien, Wiesbaden, 1990.

[MH36a] R. J. Moon and W. D. Harkins. The production of high velocity particles in a cyclotron by the use of multiphase oscillators. *Physical Review*, 49:273, 1936. Doi: 10.1103/PhysRev.49.273.2.

[MH36b] R. J. Moon and W. D. Harkins. Use of multiphase oscillators with a cyclotron (Lawrence) for the production of high-velocity particles. *Nature*, 137:316–317, 1936. Doi: 10.1038/137316b0.

[Mil72] J. D. Miller. Rowland and the nature of electric currents. *Isis*, 63:4–27, 1972.

[MKSG10] M. Mandea, M. Korte, M. Soloviev, and A. Gvishiani. Alexander von Humboldt's charts of the Earth's magnetic field: an assessment based on modern models. *History of Geo- and Space Sciences*, 1:63–76, 2010. Doi: https://www.hist-geo-space-sci.net/1/63/2010/.

[Mol72] A. P. Molella. *Philosophy and nineteenth-century German electrodynamics: the problem of atomic action at a distance.* PhD thesis, Cornell University, 1972.

[Moo04a] R. J. Moon. Proposal for a fundamental experiment in electrodynamics. *21st Century Science & Technology*, Fall:46, 2004. Available at `http://www.21sci-tech.com/Subscriptions/Archive/2004_F.pdf`.

[Moo04b] R. J. Moon. 'Space must be quantized'. *21st Century Science & Technology*, Fall:26–27, 2004. Available at `http://www.21sci-tech.com/Subscriptions/Archive/2004_F.pdf`.

[MRGL10] F. Martin-Rodriguez, G. B. Garcia, and M. A. Lires. Technological archaeology: technical description of the Gauss-Weber telegraph. In IEEE, editor, *2010 Second Region 8 IEEE Conference on the History of Communications*, Madrid, 2010. Histelcon. Doi: 10.1109/HISTELCON.2010.5735309.

[MS20] R. A. Martins and A. P. B. Silva. Joule's experiments on the heat evolved by metallic conductors of electricity. *Foundations of Science*, pages 1–77, 2020. Doi: 10.1007/s10699-020-09681-1.

[Ner33] J. J. Nervander. Mémoire sur un Galvanomètre à châssis cylindrique par lequel on obtient immédiatement et sans calcul la mesure de l'intensité du courant électrique qui produit la déviation de l'aiguille aimantée. *Annales de Chimie et de Physique*, 55:156–184, 1833.

[Neu68] C. Neumann. *Die Principien der Elektrodynamik. Eine mathematische Untersuchung*. H. Laupp, Tübingen, 1868. Reprinted in *Mathematische Annalen*, Vol. 17, pp. 400-434 (1880).

[Neu20] C. Neumann, 2020. Principles of electrodynamics. Posted in August 2020 at `www.ifi.unicamp.br/~assis`. Translated by Laurence Hecht and Urs Frauenfelder. Edited by A. K. T. Assis.

[New34] I. Newton. *Mathematical Principles of Natural Philosophy*. University of California Press, Berkeley, 1934. Cajori edition.

[New90] I. Newton. *Principia — Princípios Matemáticos de Filosofia Natural*. Nova Stella/Edusp, São Paulo, 1990. Livro I: O Movimento dos Corpos. Portuguese translation by T. Ricci, L. G. Brunet, S. T. Gehring and M. H. C. Célia.

[New99] I. Newton. *The Principia: Mathematical Principles of Natural Philosophy*. University of California Press, Berkeley, 1999. A new translation by I. B. Cohen and A. Whitman, assisted by J. Budenz.

[New08] I. Newton. *Principia — Princípios Matemáticos de Filosofia Natural*. Edusp, São Paulo, 2008. Livro II: O Movimento dos Corpos (em Meios com Resistência). Livro III: O Sistema do Mundo (Tratado Matematicamente). Portuguese translation by A. K. T. Assis.

[New10] I. Newton. *Principia — Princípios Matemáticos de Filosofia Natural*. Folha de São Paulo, São Paulo, 2010. Livro III: O Sistema do Mundo (Tratado Matematicamente). Coleção Folha de São Paulo: Livros que Mudaram o Mundo, Volume 9. Portuguese translation by A. K. T. Assis.

[Oer20a] H. C. Oersted. Expériences sur l'effet du conflict électrique sur l'aiguille aimantée. *Annales de Chimie et de Physique*, 14:417–425, 1820.

[Oer20b] H. C. Oersted. Experiments on the effect of a current of electricity on the magnetic needle. *Annals of Philosophy*, 16:273–277, 1820. Translated from a printed account drawn up in Latin by the author and transmitted by him to the Editor of the *Annals of Philosophy*.

[Oer20c] J. C. Oersted. Versuche über die Wirkung des electrischen Conflicts auf die Magnetnadel. *Annalen der Physik und Chemie*, 6:295–304, 1820. Translated by Gilbert.

[Oer65] H. C. Oersted. Experiments on the effect of a current of electricity on the magnetic needle. In R. A. R. Tricker, *Early Electrodynamics — The First Law of Circulation*, pages 113–117, New York, 1965. Pergamon. Translation from Thomson's *Annals of Philosophy*, October 1820. Translated from a printed account drawn up in Latin by the author and transmitted by him to the Editor of the *Annals of Philosophy*.

[Ole96] K. M. Olesko. Precision, tolerance, and consensus: local cultures in German and British resistance standards. In J. Z. Buchwald, editor, *Scientific Credibility and Technical Standards in 19th and Early 20th Century Germany and Britain*, pages 117–156. Springer, Dordrecht, 1996.

[OM84] F. F. Offner and E. M. McMillan. Birth of synchrotron. *Physics Today*, 37:73, 1984.

[O'R65] A. O'Rahilly. *Electromagnetic Theory — A Critical Examination of Fundamentals.* Dover, New York, 1965.

[Ørs86] H. C. Ørsted. Experiências sobre o efeito do conflito elétrico sobre a agulha magnética. *Cadernos de História e Filosofia da Ciência*, 10:115–122, 1986. Translated by R. d. A. Martins.

[Ørs98] H. C. Ørsted. Experiments on the effect of the electric conflict on the magnetic needle. In K. Jelved, A. D. Jackson, and O. Knudsen, editors, *Selected Scientific Works of Hans Christian Ørsted*, pages 413–416. Princeton University Press, Princeton, 1998. Article originally written in Latin in 1820.

[Pat17] A. M. Patterson. *A German-English Dictionary for Chemists.* John Wiley & Sons, New York, 1917.

[Pic06] F. Pichler. Das Grundgesetz der Elektrodynamik von Weber: eine übersicht zur historischen Entwicklung. In F. Pichler and M. v. Renteln, editors, *Von Newton to Gauss*, pages 203–212. Trauner, Linz, 2006. Peuerbach Symposium 2006.

[PM13] E. M. Purcell and D. J. Morin. *Electricity and Magnetism.* Cambridge University Press, Cambridge, 3rd edition, 2013.

[Pog40] J. C. Poggendorff. Ueber die galvanischen Ketten aus zwei Flüssigkeiten und zwei einander nicht berührenden Metallen. *Annalen der Physik und Chemie*, 49:31–72, 1840.

[Poi25a] Poisson. Solution d'un Problème relatif au magnétisme terrestre. *Annales de Chimie et de Physique*, 30:257–263, 1825.

[Poi25b] Poisson. Solution d'un Problème relatif au Magnétisme terrestre. *Connaissance des Tems, ou des Mouvemens Célestes, a l'usage des Astronomes et des Navigateurs, por l'An 1828*, pages 322–330, 1825. Paris.

[Pot84] A. Potier. *Collection de Mémoires relatifs a la Physique*, volume 1: *Mémoires de Coulomb*. Gauthiers-Villars, Paris, 1884.

[Pou37] Pouillet. Mémoire sur la pile de Volta et sur la loi générale de l'intensité que prennent les courrants, soit qu'ils proviennent d'un seul élément, soit qu'ils proviennnent d'une pile à grande ou à petite tension. *Comptes Rendues de l'Académie des Sciences de Paris*, 4:267–279, 1837.

[Rei77] K. Reich. *Carl Friedrich Gauß — 1777/1977*. Heinz Moos Verlag, München, 1977.

[Rei02] K. Reich. *Gauß' Werke in Kurzfassung*. Erwin Rauner, Augsburg, 2002. Algorismus, Studien zur Geschichte der Mathematik und der Naturwissenschaften, Vol. 39.

[Rei11] K. Reich. Alexander von Humboldt und Carl Friedrich Gauß als Wegbereiter der neuen Disziplin Erdmagnetismus. *International Review for Humboldtian Studies*, 12:35–55, 2011.

[Rei13] K. Reich. Die Beziehungen zwischen Kopenhagen und Göttingen auf dem Gebiet des Erdmagnetismus: Ergebnisse einer Analyse der Briefe, die "Hans Christian Oersted" mit "Carl Friedrich Gauß" und "Wilhelm Weber" wechselte. *Sudhoffs Archiv*, 97:21–38, 2013.

[Rei19] K. Reich, 2019. Die Intensität der erdmagnetischen Kraft: Nr. 15, pp. xxxix-xlii. In *Carl Friedrich Gauss — Werke — Supplement* — Band 3: Varia: 15 Abhandlungen in deutscher Übersetzung. Mit einer Einleitung, einer Bibliographie und Registern herausgegeben von Karin Reich. Hrsg. von der Akademie der Wissenschaften zu Göttingen, Universitätsverlag Göttingen 2019, S. 533-594.

[Rib14] D. Ribeiro. Wilhelm Eduard Weber. *Revista de Ciência Elementar*, 2:136, 2014. Doi: 10.24927/rce2014.136.

[Rie92] E. Riecke. *Wilhelm Weber (geb. 24. October 1804, gest. 23. Juni 1891)*. Dieterichsche Verlags-Buchhandlung, Göttingen, 1892. Rede gehalten in der öffentlichen Sitzung der K. Gesellschaft der Wissenschaften am 5. December 1891. Reprinted in Abhandlungen der Königlichen Gesellschaft der Wissenschaften zu Göttingen, Vol. 38, pp. 1-44 (1892).

[Ros56] L. Rosenfeld. The velocity of light and the evolution of electrodynamics. *Il Nuovo Cimento*, Supplement to vol. 4:1630–1669, 1956. Reprinted in Cohen, R. S. and Stachel, J. J. (eds.), Selected Papers of Léon Rosenfeld. Boston Studies in the Philosophy of Science, Vol. 21. Springer, Dordrecht, Chapter 14, pp. 134-177.

[Ros81] L. Rosenfeld. Kirchhoff, Gustav Robert. In C. C. Gillispie, editor, *Dictionary of Scientific Biography*, Vol. 7, pages 379–383, New York, 1981. Charles Scribner's Sons.

[RR12] K. Reich and E. Roussanova. *Carl Friedrich Gauß und Russland: Sein Briefwechsel mit in Russland wirkenden Wissenschaftlern.* de Gruyter, Berlin, 2012. In: Abhandlungen der Akademie der Wissenschaften zu Göttingen. Neue Folge, 16.

[RR13] K. Reich and K. Roussanova. Eine kritische Bestandsaufnahme der Werkausgabe von Carl Friedrich Gauß. *Math Semesterber*, 60:217–247, 2013. Doi: 10.1007/s00591-012-0116-3.

[RR14] K. Reich and E. Roussanova. Gauss' and Weber's "Atlas of Geomagnetism" (1840) was not the first: the history of the geomagnetic atlases. *Handbook of Geomathematics*, pages 1–32, 2014. Doi: 10.1007/978-3-642-27793-1_94-1.

[RR15] K. Reich and E. Roussanova. *Carl Friedrich Gauss und Christopher Hansteen: Der Briefwechsel beider Gelehrten im historischen Kontext.* De Gruyter, Berlin, 2015.

[RR17] K. Reich and E. Roussanova. Krakau und Göttingen: Maximilian Weisses Beziehungen zu Carl Friedrich Gauß und Wilhelm Weber. *Mitteilungen der Gauss-Gesellschaft*, 54:37–60, 2017.

[Sch40] C. F. Schönbein. Notizen über eine Volta'sche Säule von ungewöhnlicher Kraft. *Annalen der Physik und Chemie*, 49:511–514, 1840.

[Sch36] H. Schimank. Carl Friedrich Gauß und Wilhelm Weber. In W. Andreas and W. von Scholz, editors, *Die Großen Deutschen*, volume 3, pages 266–279. Neue Deutsche Biographie, 1936.

[Sch93a] W. Schreier. Die drei Brüder Weber und Gustav Theodor Fechner — Untersuchungen zur medizinischen, Psycho- und technischen Physik. *International Journal of History and Ethics of Natural Sciences, Technology and Medicine*, New Serie, Vol. 2:111–116, 1993.

[Sch93b] W. Schreier. Die Stellung der Gebrüder Weber in der Geschichte der Medizin, Naturwissenschaften und Technik. In G. Berg, W. Eisenberg, W. Hergert, and W. Schreier, editors, *II. Weber — Symposium Die Gebrüder Weber — Wegbereiter interdisziplinärer Forschung in Halle und Leipzig am 16. Oktober und 18. November 1993*, pages 1–9. Fachbereich Physik der Martin-Luther-Universität Halle-Wittenberg und Wilhelm-Weber-Gesellschaft e. V., 1993.

[Sih21] A. Sihvola. Johan Jacob Nervander and the quantification of electric current. *IEEE Antennas & Propagation Magazine*, 63:123–128, 2021. Doi: 10.1109/MAP.2020.3039803.

[TA15a] M. Tajmar and A. K. T. Assis. Gravitational induction with Weber's force. *Canadian Journal of Physics*, 93:1571–1573, 2015. Doi: 10.1139/cjp-2015-0285.

[TA15b] M. Tajmar and A. K. T. Assis. Particles with negative mass: production, properties and applications for nuclear fusion and self-acceleration. *Journal of Advanced Physics*, 4:77–82, 2015. Doi: 10.1166/jap.2015.1159.

[TA15c] M. Tajmar and A. K. T. Assis. Teilchen mit negativer Masse: Produktion, Eigenschaften und Anwendungen für die Kernfusion und die Selbst-Beschleunigung. Translated (2015) from the English by M. Pohl. Original paper: M. Tajmar and A. K. T. Assis, "Particles with negative mass: production, properties and applications for nuclear fusion and self-acceleration," Journal of Advanced Physics, Vol. 4, pp. 77-82 (2015). Available at `www.ifi.unicamp.br/~assis`, 2015.

[TA16] M. Tajmar and A. K. T. Assis. Influence of rotation on the weight of gyroscopes as an explanation for flyby anomalies. *Journal of Advanced Physics*, 5:176–179, 2016. Doi: 10.1166/jap.2016.1233.

[Ten96] J. Tennenbaum. An introduction to 'The significance of the 1845 Gauss-Weber correspondence'. *21st Century Science & Technology*, 9(3):2–5, 1996.

[Ten97a] J. Tennenbaum. Demonstrating Gauss and Weber's magnetometer. *21st Century Science & Technology*, 10(1):61–62, 1997.

[Ten97b] J. Tennenbaum. Wilhelm Weber's 'Aphorisms' — On the hypotheses of temporal ordering. *21st Century Science & Technology*, 10(2):50–51, 1997.

[Tho74] W. Thomson *et al.* First report of the committee for the selection and nomenclature of dynamical and electrical units. *Report of the British Association for the Advancement of Science (B. A. Report)*, pages 222–225, 1874.

[Tim05] A. Timm. Der elektromagnetische Telegraph von Gauß und Weber. *Göttinger Bibliothekschriften*, 30:169–187, 2005.

[Tom20] F. D. Tombe. The 1855 Weber-Kohlrausch experiment (the speed of light). *The General Science Journal*, 2020. Available at `www.gsjournal.net/Science-Journals/Research%20Papers/View/7711`.

[Web37a] W. Weber. Bemerkungen über die Einrichtung magnetischer Observatorien und Beschreibung der darin aufzustellenden Instrumente. In C. F. Gauss and W. Weber, editors, *Resultate aus den Beobachtungen des magnetischen Vereins im Jahre 1836*, volume I, chapter I, pages 13–33. Dieterichschen Buchhandlung, Göttingen, 1837. Reprinted in Wilhelm Weber's *Werke*, Vol. 2, E. Riecke (ed.), (Springer, Berlin, 1892), pp. 3-19.

[Web37b] W. Weber. Beschreibung eines kleinen Apparats zur Messung des Erdmagnetismus nach absolutem Maass für Reisende. In C. F. Gauss and W. Weber, editors, *Resultate aus den Beobachtungen des magnetischen Vereins im Jahre 1836*, volume I, chapter IV, pages 63–89. Dieterichschen Buchhandlung, Göttingen, 1837. Reprinted in Wilhelm Weber's *Werke*, Vol. 2, E. Riecke (ed.), (Springer, Berlin, 1892), pp. 20-42.

[Web38a] W. Weber. Bemerkungen über die Einrichtung und den Gebrauch des Bifilar-Magnetometers. In C. F. Gauss and W. Weber, editors, *Resultate aus den Beobachtungen des magnetischen Vereins im Jahre 1837*, volume II, chapter II, pages 20–37. Dieterichschen Buchhandlung, Göttingen, 1838. Reprinted in Wilhelm Weber's *Werke*, Vol. 2, E. Riecke (ed.), (Springer, Berlin, 1892), pp. 43-57.

[Web38b] W. Weber. Description d'un petit appareil portatif pour les voyageurs, et destiné aux mesures absolues du magnétisme terrestre. In A. Quetelet, editor, *Correspondance Mathématique et Physique*, pages 308–338. Société Belge de Librairie, Bruxelles, 1838. Tome X et Tome Second de la Troisième Série.

[Web38c] W. Weber. Remarques sur l'établissement des observatoires magnétiques, et description des instrumens à y placer. In A. Quetelet, editor, *Correspondance Mathématique et Physique*, pages 49–71. Société Belge de Librairie, Bruxelles, 1838. Tome X et Tome Second de la Troisième Série.

[Web39] W. Weber. Das transportable Magnetometer. In C. F. Gauss and W. Weber, editors, *Resultate aus den Beobachtungen des magnetischen Vereins im Jahre 1838*, volume III, chapter III, pages 68–85. Weidmannschen Buchhandlung, Leipzig, 1839. Reprinted in Wilhelm Weber's *Werke*, Vol. 2, E. Riecke (ed.), (Springer, Berlin, 1892), pp. 89-104.

[Web40] W. Weber. Erläuterungen zu den Terminszeichnungen und den Beobachtungszahlen. In C. F. Gauss and W. Weber, editors, *Resultate aus den Beobachtungen des magnetischen Vereins im Jahre 1839*, volume IV, chapter VIII, pages 120–130. Weidmannschen Buchhandlung, Leipzig, 1840. Reprinted in Wilhelm Weber's *Werke*, Vol. 2, E. Riecke (ed.), (Springer, Berlin, 1892), pp. 182-189.

[Web41a] W. Weber. Description of a small portable apparatus for measuring the absolute intensity of terrestrial magnetism. In R. Taylor, editor, *Scientific Memoirs,* Vol. 2, pages 65–87, London, 1841. Taylor and Francis. From the *Resultate aus den Beobachtungen des magnetischen Vereins im Jahre 1836*, Volume I, Chapter IV, pp. 63-89. Translated by Mr. William Francis, and revised by Professor Lloyd and Major Sabine. Available at `https://www.biodiversitylibrary.org/bibliography/2501#/summary`.

[Web41b] W. Weber. An extract from remarks on the term-observations for 1839, of the German Magnetic Association. In R. Taylor, editor, *Scientific Memoirs,* Vol. 2, pages 587–588, London, 1841. Taylor and Francis. From the *Resultate aus den Beobachtungen des magnetischen Vereins im Jahre 1839*, Volume IV, Chapter VIII, pp. 120-130. Available at `https://www.biodiversitylibrary.org/bibliography/2501#/summary`.

[Web41c] W. Weber. Messung starker galvanischer Ströme bei geringem Widerstande nach absolutem Maasse. In C. F. Gauss and W. Weber, editors, *Resultate aus den Beobachtungen des magnetischen Vereins im Jahre 1840*, volume V, chapter VIII, pages 83–90. Weidmannschen Buchhandlung, Leipzig, 1841. Reprinted in Wilhelm Weber's *Werke*, Vol. 3, H. Weber (ed.), (Springer, Berlin, 1893), pp. 6-12.

[Web41d] W. Weber. Observations on the arrangement and use of the bifilar magnetometer. In R. Taylor, editor, *Scientific Memoirs,* Vol. 2, pages 268–280, London, 1841. Taylor and Francis. From the *Resultate aus den Beobachtungen des magnetischen Vereins im Jahre 1837*, Volume II, Chapter II, pp. 20-37.—Herausgegeben von Carl Friedrich Gauss und Wilhelm Weber. Göttingen, 1838. Translated by Mr. William Francis, and revised by Professor Lloyd and Major Sabine. Available at `https://www.biodiversitylibrary.org/bibliography/2501#/summary`.

[Web41e] W. Weber. On a transportable magnetometer. In R. Taylor, editor, *Scientific Memoirs*, Vol. 2, pages 565–586, London, 1841. Taylor and Francis. This article is translated partly from the *Resultate aus den Beobachtungen des magnetischen Vereins im Jare 1838*, Volume III, Chapter III, pp. 68-85, and partly from manuscript communications from M. Weber to Major Sabine. Translation presented by Major Sabine. Available at `https://www.biodiversitylibrary.org/bibliography/2501#/summary`.

[Web41f] W. Weber. Remarks on the arrangement of magnetical observatories, and description of the instruments to be placed in them. In R. Taylor, editor, *Scientific Memoirs*, Vol. 2, pages 25–42, London, 1841. Taylor and Francis. From the *Resultate aus den Beobachtungen des magnetischen Vereins im Jahre 1836*, Volume I, Chapter I, pp. 13-33. Translated by Mr. William Francis, and revised by Professor Lloyd and Major Sabine. Available at `https://www.biodiversitylibrary.org/bibliography/2501#/summary`.

[Web41g] W. Weber. Über magnetische Friction. In C. F. Gauss and W. Weber, editors, *Resultate aus den Beobachtungen des magnetischen Vereins im Jahre 1840*, volume V, chapter IV, pages 46–58. Weidmannschen Buchhandlung, Leipzig, 1841. Reprinted in Wilhelm Weber's *Werke*, Vol. 2, E. Riecke (ed.), (Springer, Berlin, 1892), pp. 200-211.

[Web41h] W. Weber. Ueber das elektrochemische Aequivalent des Wassers. In C. F. Gauss and W. Weber, editors, *Resultate aus den Beobachtungen des magnetischen Vereins im Jahre 1840*, volume V, chapter IX, pages 91–98. Weidmannschen Buchhandlung, Leipzig, 1841. Reprinted in Wilhelm Weber's *Werke*, Vol. 3, H. Weber (ed.), (Springer, Berlin, 1893), pp. 13-18.

[Web42a] W. Weber. Bemerkungen über die Wirkungen eines Magnetes in die Ferne. *Annalen der Physik und Chemie*, 55:33–42, 1842. Reprinted in Wilhelm Weber's *Werke*, Vol. 2, E. Riecke (ed.), (Springer, Berlin, 1892), pp. 242-249.

[Web42b] W. Weber. Messung starker galvanischer Ströme nach absolutem Maasse. *Annalen der Physik und Chemie*, 55:27–32, 1842. Reprinted in Wilhelm Weber's *Werke*, Vol. 3, H. Weber (ed.), (Springer, Berlin, 1893), pp. 19-23.

[Web42c] W. Weber. Über ein neues Instrument zur Beobachtung galvanischer Ströme. *Amtlicher Bericht über die neunzehnte Versammlung deutscher Naturforscher und Ärzte zu Braunschweig im September 1841*, pages 154–158, 1842. Available at `https://opacplus.bsb-muenchen.de/title/11134128`.

[Web42d] W. Weber. Ueber das elektrochemische Aequivalent des Wassers. *Annalen der Physik und Chemie*, 55:181–189, 1842. Reprinted in Wilhelm Weber's *Werke*, Vol. 3, H. Weber (ed.), (Springer, Berlin, 1893), pp. 13-18.

[Web46] W. Weber. Elektrodynamische Maassbestimmungen — Über ein allgemeines Grundgesetz der elektrischen Wirkung. *Abhandlungen bei Begründung der Königlich Sächsischen Gesellschaft der Wissenschaften am Tage der zweihundertjährigen Geburtstagfeier Leibnizen's herausgegeben von der Fürstlich Jablonowskischen Gesellschaft (Leipzig)*, pages 211–378, 1846. Reprinted in Wilhelm Weber's *Werke*, Vol. 3, H. Weber (ed.), (Springer, Berlin, 1893), pp. 25-214.

[Web48a] W. Weber. Elektrodynamische Maassbestimmungen. *Annalen der Physik und Chemie*, 73:193–240, 1848. Reprinted in Wilhelm Weber's *Werke*, Vol. 3, H. Weber (ed.), (Springer, Berlin, 1893), pp. 215-254.

[Web48b] W. Weber. Über die Erregung und Wirkung des Diamagnetismus nach den Gesetzen inducirter Ströme. *Annalen der Physik und Chemie*, 73:241–256, 1848. Reprinted in Wilhelm Weber's *Werke*, Vol. 3, H. Weber (ed.), (Springer, Berlin, 1893), pp. 255-268.

[Web51] W. Weber. Messungen galvanischer Leitungswiderstände nach einem absolutem Maasse. *Annalen der Physik und Chemie*, 82:337–369, 1851. Reprinted in Wilhelm Weber's *Werke*, Vol. 3, H. Weber (ed.), (Springer, Berlin, 1893), pp. 276-300.

[Web52a] W. Weber. Elektrodynamische Maassbestimmungen insbesondere über Diamagnetismus. *Abhandlungen der Königlich Sächsischen Gesellschaft der Wissenschaften zu Leipzig, mathematisch-physischen Classe*, 1:485–577, 1852. Reprinted in Wilhelm Weber's *Werke*, Vol. 3, H. Weber (ed.), (Springer, Berlin, 1893), pp. 473-554.

[Web52b] W. Weber. Elektrodynamische Maassbestimmungen insbesondere Widerstandsmessungen. *Abhandlungen der Königlich Sächsischen Gesellschaft der Wissenschaften zu Leipzig, mathematisch-physischen Classe*, 1:199–381, 1852. Reprinted in Wilhelm Weber's *Werke*, Vol. 3, H. Weber (ed.), (Springer, Berlin, 1893), pp. 301-471.

[Web52c] W. Weber. On the excitation and action of diamagnetism according to the laws of induced currents. In R. Taylor, editor, *Scientific Memoirs,* Vol. 5, pages 477–488, London, 1852. Taylor and Francis. Translated by W. Francis from Poggendorff's *Annalen*, Jan. 7, 1848. Available at `https://archive.org/details/in.ernet.dli.2015.212784` and `https://www.biodiversitylibrary.org/bibliography/2501#/summary`.

[Web52d] W. Weber. On the measurement of electro-dynamic forces. In R. Taylor, editor, *Scientific Memoirs,* Vol. 5, pages 489–529, London, 1852. Taylor and Francis. Available at `https://archive.org/details/in.ernet.dli.2015.212784` and `https://www.biodiversitylibrary.org/bibliography/2501#/summary`.

[Web52e] W. Weber. Ueber den Zusammenhang der Lehre vom Diamagnetismus mit der Lehre von dem Magnetismus und der Elektricität. *Annalen der Physik und Chemie*, 87:145–189, 1852. Reprinted in Wilhelm Weber's *Werke*, Vol. 3, H. Weber (ed.), (Springer, Berlin, 1893), pp. 555-590.

[Web53] W. Weber. On the connexion of diamagnetism with magnetism and electricity. In J. Tyndall and W. Francis, editors, *Scientific Memoirs,* Vol. on Natural Philosophy, pages 163–199, London, 1853. Taylor and Francis. Translated by J. Tyndall. Available at `https://books.google.com.br/books?id=C1i4AAAAIAAJ&hl=pt-BR&source=gbs_navlinks_s`.

[Web55] W. Weber. On the theory of diamagnetism — Letter from Professor Weber to Prof. Tyndall. *Philosophical Magazine*, 10:407–410, 1855. Translated by J. Tyndall. Reprinted in J. Tyndall, Researches on Diamagnetism and Magne-crystallic Action including the Question of Diamagnetic Polarity (D. Appleton and Company, New York, 1888), pp. 243-249.

[Web61] W. Weber. On the measurement of electric resistance according to an absolute standard. *Philosophical Magazine*, 22:226–240 and 261–269, 1861. Translated by Dr. E. Atkinson.

[Web62] W. Weber. Zur Galvanometrie. *Abhandlungen der Königlich Gesellschaft der Wissenschaften zu Göttingen, mathematische Klasse*, 10:3–96, 1862. Reprinted in Wilhelm Weber's *Werke*, Vol. 4, H. Weber (ed.), (Springer, Berlin, 1894), pp. 17-96.

[Web64] W. Weber. Elektrodynamische Maassbestimmungen insbesondere über elektrische Schwingungen. *Abhandlungen der Königlich Sächsischen Gesellschaft der Wissenschaften zu Leipzig, mathematisch-physischen Classe*, 6:571–716, 1864. Reprinted in Wilhelm Weber's *Werke*, Vol. 4, H. Weber (ed.), (Springer, Berlin, 1894), pp. 105-241.

[Web71] W. Weber. Elektrodynamische Maassbestimmungen insbesondere über das Princip der Erhaltung der Energie. *Abhandlungen der Königlich Sächsischen Gesellschaft der Wissenschaften zu Leipzig, mathematisch-physischen Classe*, 10:1–61, 1871. Reprinted in Wilhelm Weber's *Werke*, Vol. 4, H. Weber (ed.), (Springer, Berlin, 1894), pp. 247-299.

[Web72] W. Weber. Electrodynamic measurements — Sixth memoir, relating specially to the principle of the conservation of energy. *Philosophical Magazine*, 43:1–20 and 119–149, 1872. Translated by Professor G. C. Foster, F.R.S., from the *Abhandlungen der mathem.-phys. Classe der Königlich Sächsischen Gesellschaft der Wissenschaften*, vol. x (January 1871).

[Web78] W. Weber. Elektrodynamische Maassbestimmungen insbesondere über die Energie der Wechselwirkung. *Abhandlungen der Königlich Sächsischen Gesellschaft der Wissenschaften zu Leipzig, mathematisch-physischen Classe*, 11:641–696, 1878. Reprinted in Wilhelm Weber's *Werke*, Vol. 4, H. Weber (ed.), (Springer, Berlin, 1894), pp. 361-412.

[Web87] W. Weber. Mesures électrodynamiques. In J. Joubert, editor, *Collection de Mémoires relatifs a la Physique,* Vol. III: *Mémoires sur l'Électrodynamique*, pages 289–402. Gauthier-Villars, Paris, 1887.

[Web92a] H. Weber. Wilhelm Weber — Eine Lebensskizze. *Deutsche Revue*, 3:183–202, 1892.

[Web92b] H. Weber. Wilhelm Weber — Eine Lebensskizze (Fortsetzung). *Deutsche Revue*, 3:324–345, 1892.

[Web92c] H. Weber. Wilhelm Weber — Eine Lebensskizze (Schluß). *Deutsche Revue*, 4:99–118, 1892.

[Web92d] W. Weber. *Wilhelm Weber's Werke,* E. Riecke (ed.), volume 2, *Magnetismus*. Springer, Berlin, 1892. Available at `http://archive.org/details/wilhelmweberswe01fiscgoog`.

[Web92e] W. Weber. *Wilhelm Weber's Werke,* W. Voigt, (ed.), volume 1, *Akustik, Mechanik, Optik und Wärmelehre*. Springer, Berlin, 1892.

[Web93a] H. Weber. *Wilhelm Weber — Eine Lebensskizze.* Trewendt, Breslau, 1893.

[Web93b] W. Weber. *Wilhelm Weber's Werke,* H. Weber (ed.), volume 3, *Galvanismus und Elektrodynamik*, first part. Springer, Berlin, 1893. Available at `http://archive.org/details/wilhelmweberswe02fiscgoog`.

[Web94a] W. Weber. Aphorismen. In H. Weber, editor, Wilhelm Weber's *Werke,* Vol. 4, pages 630–632, Berlin, 1894. Springer.

[Web94b] W. Weber. Elektrodynamische Maassbestimmungen insbesondere über den Zusammenhang des elektrischen Grundgesetzes mit dem Gravitationsgesetze. In H. Weber, editor, Wilhelm Weber's *Werke,* Vol. 4, pages 479–525, Berlin, 1894. Springer.

[Web94c] W. Weber. Ueber die Einrichtung des Bifilargalvanometers. In H. Weber, editor, Wilhelm Weber's *Werke,* Vol. 4, pages 584–615, Berlin, 1894. Springer.

[Web94d] W. Weber. *Wilhelm Weber's Werke,* H. Weber, (ed.), volume 4, *Galvanismus und Elektrodynamik*, second part. Springer, Berlin, 1894.

[Web66a] W. Weber. An extract from remarks on the term-observations for 1839, of the German Magnetic Association. In R. Taylor, editor, *Scientific Memoirs,* Vol. 2, pages 587–588, New York, 1966. Johnson Reprint Corporation.

[Web66b] W. Weber. Observations on the arrangement and use of the bifilar magnetometer. In R. Taylor, editor, *Scientific Memoirs,* Vol. 2, pages 268–280, New York, 1966. Johnson Reprint Corporation.

[Web66c] W. Weber. On a transportable magnetometer. In R. Taylor, editor, *Scientific Memoirs,* Vol. 2, pages 565–586, New York, 1966. Johnson Reprint Corporation.

[Web66d] W. Weber. On the connexion of diamagnetism with magnetism and electricity. In J. Tyndall and W. Francis, editors, *Scientific Memoirs,* Vol. on Natural Philosophy, pages 163–199, New York, 1966. Johnson Reprint Corporation. Translated by J. Tyndall.

[Web66e] W. Weber. On the excitation and action of diamagnetism according to the laws of induced currents. In R. Taylor, editor, *Scientific Memoirs,* Vol. 5, pages 477–488, New York, 1966. Johnson Reprint Corporation. Translated by W. Francis.

[Web66f] W. Weber. On the measurement of electro-dynamic forces. In R. Taylor, editor, *Scientific Memoirs,* Vol. 5, pages 489–529, New York, 1966. Johnson Reprint Corporation.

[Web97] W. Weber. Aphorisms. *21st Century Science & Technology*, 10(2):52–53, 1997. English translation by J. Tennenbaum.

[Web07] W. Weber, 2007. Determinations of electrodynamic measure: concerning a universal law of electrical action, 21st Century Science & Technology, posted March 2007, translated by S. P. Johnson, edited by L. Hecht and A. K. T. Assis. Available at `http://21sci-tech.com/translation.html` and `www.ifi.unicamp.br/~assis`.

[Web08] W. Weber, 2008. Determinations of electrodynamic measure: particularly in respect to the connection of the fundamental laws of electricity with the law of gravitation, 21st Century Science & Technology, posted November 2008, translated by G. Gregory, edited by L. Hecht and A. K. T. Assis. Available at `http://21sci-tech.com/translation.html` and `www.ifi.unicamp.br/~assis`.

[Web19] W. Weber, 2019. On the measurement of electro-dynamic forces. Classics of Measure no. 1. Second edition. Version of 21 February 2019. This English translation appeared originally in R. Taylor (editor), *Scientific Memoirs, selected from the Transactions of Foreign Academies of Science and Learned Societies*, Volume V, part 20, article 14 (London: Taylor and Francis, 1852). Available at `www.sizes.com/library/classics/Weber1.pdf`.

[Web20a] W. Weber, 2020. Measurement of strong galvanic currents with low resistance according to absolute measure. Posted in July 2020 at `www.ifi.unicamp.br/~assis`. Translated by H. Härtel. Edited by A. K. T. Assis.

[Web20b] W. Weber, 2020. On the electrochemical equivalent of water. Posted in April 2020 at `www.ifi.unicamp.br/~assis`. Translated by H. Härtel. Edited by A. K. T. Assis.

[Web20c] W. Weber, 2020. Measurement of strong galvanic currents according to absolute measure. Posted in October 2020 at `www.ifi.unicamp.br/~assis`. Translated by H. Härtel. Edited by A. K. T. Assis.

[Web21a] W. Weber, 2021. Electrodynamic measurements, especially on electric oscillations. Third version posted in February 2021 at `www.ifi.unicamp.br/~assis`. Translated by P. Marquardt and edited by A. K. T. Assis.

[Web21b] W. Weber, 2021. Electrodynamic measurements, especially measures of resistance. Second version posted in February 2021 at `www.ifi.unicamp.br/~assis`. Translated by D. H. Delphenich and edited by A. K. T. Assis.

[Web21c] W. Weber, 2021. Electrodynamic measurements, especially on diamagnetism. Third version posted in February 2021 at `www.ifi.unicamp.br/~assis`. Translated by U. Frauenfelder. Edited by U. Frauenfelder and A. K. T. Assis.

[Web21d] W. Weber, 2021. Electrodynamic measurements, especially on the energy of interaction. Fourth version posted in February 2021 at `www.ifi.unicamp.br/~assis`. Translated by Joa Weber. Edited by A. K. T. Assis.

[Wes91] J. P. Wesley. *Selected Topics in Advanced Fundamental Physics*. Benjamin Wesley Publisher, Blumberg, 1991.

[Wes01] J. P. Wesley. The Weber cosmological condition and Wesley gravitation. *Physics Essays*, 14:160–163, 2001.

[Whi73] E. T. Whittaker. *A History of the Theories of Aether and Electricity*, volume 1: *The Classical Theories*. Humanities Press, New York, 1973.

[Wie60] K. H. Wiederkehr. Wilhelm Webers Stellung in der Entwicklung der Elektrizitätslehre. Dissertation, Hamburg, 1960.

[Wie64] K. H. Wiederkehr. Aus der Geschichte des Göttinger Magnetischen Vereins und seiner Resultate. *Nachrichten der Akademie der Wissenschaften in Göttingen, II. Mathematisch-Physikalische Klasse*, 14:165–205, 1964.

[Wie67] K. H. Wiederkehr. *Wilhelm Eduard Weber — Erforscher der Wellenbewegung und der Elektrizität (1804-1891)*, volume 32 of *Grosse Naturforscher*, H. Degen (ed.). Wissenschaftliche Verlagsgesellschaft, Stuttgart, 1967.

[Wie73] K. H. Wiederkehr. Das bisher unbekannte Gauß-Gutachten zur Wiederbesetzung des Göttinger Physiklehrstuhls 1831. *Gauß-Gesellschaft, Mitteilungen*, 10:32–47, 1973.

[Wie82] K. H. Wiederkehr. Über die Verleihung der Copley-Medaille an Gauß und die Mitarbeit Englands im Göttinger Magnetischen Verein. *Gauß-Gesellschaft, Mitteilungen*, 19:15–35, 1982.

[Wie88] K. H. Wiederkehr. Zur Deutung magnetischer Phänomene im 19. Jahrhundert. *Physikalische Blätter*, 44:129–134, 1988.

[Wie90] K. H. Wiederkehr. Freunde und Förderer Bernhard Riemanns, Zwei Briefe aus seinen letzten Lebensjahren. *Gauß-Gesellschaft, Mitteilungen*, 27:75–85, 1990.

[Wie91] K. H. Wiederkehr. Faradays Feldkonzept und Hans Christian Oersted. *Physikalische Blätter*, 47:825–830, 1991.

[Wie92] K. H. Wiederkehr. Wilhelm Weber und die Entwicklung in der Geomagnetik und Elektrodynamik. *Gauß-Gesellschaft, Mitteilungen*, 29:63–72, 1992.

[Wie93a] K. H. Wiederkehr. Das Experiment von Wilhelm Weber und Rudolf Kohlrausch 1855 und Maxwells elektromagnetische Lichttheorie. In W. Schröder, editor, *The Earth and the Universe (A Festschrift in honour of Hans-Jürgen Treder)*, pages 452–463. Interdivisional Commission on History of the International Association of Geomagnetism and Aeronomy, Bremen-Rönnebeck, 1993.

[Wie93b] K. H. Wiederkehr. W. Weber und die Entwicklung der Elektrodynamik. In G. Berg, W. Eisenberg, W. Hergert, and W. Schreier, editors, *II. Weber — Symposium Die Gebrüder Weber — Wegbereiter interdisziplinärer Forshcung in Halle und Leipzig am 16. Oktober und 18. November 1993*, pages 39–54. Fachbereich Physik der Martin-Luther-Universität Halle-Wittenberg und Wilhelm-Weber-Gesellschaft e. V., 1993.

[Wie93c] K. H. Wiederkehr. Wilhelm Weber und die Entwicklung in der Geomagnetik und Elektrodynamik. In G. Berg, W. Eisenberg, W. Hergert, and H. Thieme, editors, *I. Weber — Symposium anläßlich des 100. Todestages W. Webers in Halle und Wittenberg am 20. und 21. Juni 1991*, pages 1–14. Fachbereich Physik der Martin-Luther-Universität Halle-Wittenberg und Wilhelm-Weber-Gesellschaft e. V., 1993.

[Wie94] K. H. Wiederkehr. Wilhelm Weber und Maxwells elektromagnetische Lichttheorie. *Gesnerus, Part. 3/4*, 51:256–267, 1994.

[Wie97] K. H. Wiederkehr. Carl Friedrich Gauss (1777-1855) und Wilhelm Weber (1804-1891). In K. v. Meÿenn, editor, *Die Grossen Physiker*, Vol. 1, pages 357–370 and 522–524. Verlag C. H. Beck, München, 1997.

[Wie04] K. H. Wiederkehr. Ein bisher unveröffentlichter Brief von Rudolf Kohlrausch an Andreas v. Ettingshausen von 1854, das Kohlrausch-Weber-Experiment von 1854/55 und die Lichtgeschwindigkeit in Wilhelm Webers Elektrodynamik. *NTM International Journal of History and Ethics of Natural Sciences, Technology and Medicine*, 12:129–145, 2004.

[Wie07] K. H. Wiederkehr. Über Vorstellungen vom Wesen des elektrischen Stromes bis zum Beginn der Elektronentheorie der Metalle. In G. Wolfschmidt, editor, *"Es gibt für Könige keinen besonderen Weg zur Geometrie" — Festschrift für Karin Reich*, pages 299–308. Dr. Erwin Rauner Verlag, Augsburg, 2007.

[Wie08] K. H. Wiederkehr. Heinrich Hertz between the old electrodynamics and Maxwell's theory. In G. Wolfschmidt, editor, *Heinrich Hertz (1857-1894) and the Development of Communication*, pages 151–159. Schwerpunkt Geschichte der Naturwissenschaften, Mathematik und Technik, Hamburg, 2008. Nuncius Hamburgensis: Beiträge zur Geschichte der Naturwissenschaften, Vol. 10.

[Wis81] M. N. Wise. German concepts of force, energy, and the electromagnetic ether: 1845–1880. In G. N. Cantor and M. J. S. Hodge, editors, *Conceptions of Ether — Studies in the History of Ether Theories 1740–1900*, pages 269–307, Cambridge, 1981. Cambridge University Press.

[Wis21] M. N. Wise. Does narrative matter? Engendering belief in electromagnetic theory. In M. Carrier, R. Mertens, and C. Reinhardt, editors, *Narratives and Comparisons: Adversaries or Allies in Understanding Science?*, pages 29–62. Bielefeld University Press, Bielefeld, 2021.

[WK56] W. Weber and R. Kohlrausch. Über die Elektricitätsmenge, welche bei galvanischen Strömen durch den Querschnitt der Kette fliesst. *Annalen der Physik und Chemie,* J. C. Poggendoff (ed.), 99:10–25, 1856. Reprinted in Wilhelm Weber's *Werke*, Vol. 3, H. Weber (ed.), (Springer, Berlin, 1893), pp. 597-608.

[WK68] W. Weber and R. Kohlrausch. Über die Einführung absoluter elektrischer Maße. In S. Balke, H. Gericke, W. Hartner, G. Kerstein, F. Klemm, A. Portmann, H. Schimank, and K. Vogel, editors, *Ostwalds Klassiker der exakten Wissenschaften,* new series, Vol. 5. Friedrich-Vieweg & Sohn, Braunschweig, 1968. Commented by F. Kohlrausch and K. H. Wiederkehr.

[WK03] W. Weber and R. Kohlrausch. On the amount of electricity which flows through the cross-section of the circuit in galvanic currents. In F. Bevilacqua and E. A. Giannetto, editors, *Volta and the History of Electricity*, pages 287–297. Università degli Studi di Pavia and Editore Ulrico Hoepli, Milano, 2003. Translated by S. P. Johnson and edited by L. Hecht. Available at `http://ppp.unipv.it/PagesIT/VoltaHistElecSec3Frame.htm` and `www.ifi.unicamp.br/~assis`.

[WK08] W. Weber and R. Kohlrausch. Sobre a quantidade de eletricidade que flui através da seção reta do circuito em correntes galvânicas. *Revista Brasileira de História da Ciência*, 1:94–102, 2008. Portuguese translation by A. K. T. Assis.

[WM06] J. P. Wesley and P. Marquardt. *Light a Photon Flux and Other Topics*. Benjamin Wesley Publisher, Blumberg, 2006.

[Wol05] G. Wolfschmidt (ed.). *Vom Magnetismus zur Elektrodynamik: anläßlich des 200. Geburtstags von Wilhelm Weber (1804-1891) und des 150. Todestages von Carl Friedrich Gauss (1777-1855)*. Schwerpunkt Geschichte der Naturwissenschaften, Mathematik und Technik, Hamburg, 2005.

[Woo62] A. E. Woodruff. Action at a distance in nineteenth century electrodynamics. *Isis*, 53:439–459, 1962.

[Woo68] A. E. Woodruff. The contributions of Hermann von Helmholtz to electrodynamics. *Isis*, 59:300–311, 1968.

[Woo81] A. E. Woodruff. Weber, Wilhelm Eduard. In C. C. Gillispie, editor, *Dictionary of Scientific Biography,* Vol. 14, pages 203–209, New York, 1981. Charles Scribner's Sons.

[WW41a] F. Wöhler and W. Weber. Neue galvanische Säule. *Journal für praktische Chemie*, 23:313–316, 1841.

[WW41b] F. Wöhler and W. Weber. Ueber eine neue Construction der Grove'schen Säulen. *Annalen der Chemie und Pharmacie*, 38:307–311, 1841. Reprinted in *Polytechnisches Journal*, Vol. 81, pp. 273-275 (1841).

[WW41c] F. Wöhler and W. Weber. Zusammensetzung galvanischer Säulen. *Göttingische gelehrte Anzeigen*, 1:801–804, 1841. 81. Stuck der *Göttingische gelehrte Anzeigen*, den 24. May 1841. Reprinted in Wilhelm Weber's *Werke*, Vol. 3, H. Weber (ed.), (Springer, Berlin, 1893), pp. 3-5.

[WW93] E. H. Weber and W. Weber. *Wilhelm Weber's Werke,* E. Riecke (ed.), volume 5, *Wellenlehre auf Experimente gegründet oder über die Wellen tropfbarer Flüssigkeiten mit Anwendung auf die Schall- und Lichtwellen.* Springer, Berlin, 1893. Originally published in 1825. Available at `http://archive.org/details/wilhelmweberswe00fiscgoog`.

[WW94] W. Weber and E. Weber. *Wilhelm Weber's Werke,* F. Merkel and O. Fischer (editors), volume 6, *Mechanik der menschlichen Gehwerkzeuge. Eine anatomisch-physiologische Untersuchung.* Springer, Berlin, 1894. Originally published in 1836. Available at `http://archive.org/details/wilhelmweberswe03fiscgoog`.

[WW92] W. Weber and E. Weber. *Mechanics of the Human Walking Apparatus*. Springer, Berlin, 1992. Translated by P. Maquet and R. Furlong.

[WW20] F. Wöhler and W. Weber, 2020. Composition of galvanic piles. Posted in August 2020 at `www.ifi.unicamp.br/~assis`. Translated by H. Härtel and edited by A. K. T. Assis.

[Zöl78] F. Zöllner. *Wissenschaftliche Abhandlungen*, volume 2, Part I. L. Staackmann, Leipzig, 1878.

Made in the USA
Monee, IL
05 November 2026

67381966R00143